Transport and Energy Processes 2018

AA009128

Core Programming Area at the 2018 AIChE Annual Meeting

Pittsburgh, Pennsylvania, USA
28 October - 2 November 2018

ISBN: 978-1-5108-7626-2

Printed from e-media with permission by:

Curran Associates, Inc.
57 Morehouse Lane
Red Hook, NY 12571

Printed by Curran Associates, Inc. (2019)

For permission requests, please contact AIChE
at the address below.

AIChE
120 Wall Street, FL 23
New York, NY 10005-4020

Phone: (800) 242-4363
Fax: (203) 775-5177

www.aiche.org

Additional copies of this publication are available from:

Curran Associates, Inc.
57 Morehouse Lane
Red Hook, NY 12571 USA
Phone: 845-758-0400
Fax: 845-758-2633
Email: curran@proceedings.com
Web: www.proceedings.com

Transport and Energy Processes 2018

Core Programming Area at the 2018 AIChE Annual Meeting

Pittsburgh, Pennsylvania, USA
28 October - 2 November 2018

AIChE
The Global Home of Chemical Engineers

TABLE OF CONTENTS

Some abstracts in this book refer to images that were not made available by the publisher

537092 Modeling the Effect of Mesoporous Carbon Cathodes in Li-S Batteries

Abstract:

Lithium sulfur (Li-S) batteries are a promising candidate for next-generation energy storage due to their high theoretical energy density. However, well-known challenges such as polysulfide crossover, large volume changes during cycling, and passivation of reaction sites limit the capacity and capacity retention of Li-S batteries. One approach to address these issues is to control the size of pores in the carbon cathode, which affects the transport of sulfur, surface area for reaction, and lithium sulfide deposition during cycling. In particular, mesopores (2-50nm diameter) have shown prior success in Li-S cathode materials due to their high surface area and their ability to retain polysulfide intermediates while still maintaining electrolyte access.

To investigate the effect of mesopores on Li-S battery performance, we have developed a 1-dimension + time numerical continuum model for the Li-S full cell. The model includes the electrochemical reactions and precipitation of sulfur and intermediate polysulfide species, multicomponent transport of species through the pore, polysulfide crossover, lithium sulfide nucleation, and passivation of the cathode. We use the model to predict optimal pore size distributions for different charge rates and sulfur loading, which are compared to results from experimental Li-S coin cells made with mesoporous carbon nanofiber cathodes.

THEORETICAL AND EXPERIMENTAL CHARACTERIZATIONS OF A RECHARGEABLE HYBRID CATHODE FOR LITHIUM-BASED BATTERIES

Sarwan S. Sandhu and Clayton J. Cashion, Department of Chemical and Materials Engineering, University of Dayton, Dayton, OH 45469

Joseph P. Fellner, Air Force Research Laboratory, Wright-Patterson Air Force Base, OH 45433

Abstract

Theoretical and experimental investigations on a hybrid cathode for use in a potentially rechargeable lithium-based cell have been performed. The hybrid cathode active material consists of a mixture of carbon monofluoride (CF_x) and copper phthalocyanine (CuPc). The CF_x is used on the first cycle to form a highly conducting graphene-like structure to enhance the cathode electronic conductivity. We will present experimental cell discharge data using this new and innovative hybrid concept, model formulation for the charge transfer electro-kinetics for the cathode electrolyte/active material, and computational results on the structure/composition of the CuPc active material at different states of charge.

1.0 Introduction

Our understanding of the experimental data [1-3] on the electrochemical characterization of a cathode active material such as copper phthalocyanine ($CuPc_{(s)}$) or iron phthalocyanine ($FePc_{(s)}$) in the lithium-based cell: $Li_{(s)}$/electrolyte/cathode active material $CuPc_{(s)}$ or $FePc_{(s)}$, has led us to state that the active materials $CuPc_{(s)}$ or $FePc_{(s)}$ act as the host materials. For every lithium ion entering the cell cathode active material across the interface between the cell electrolyte and the cathode active material, for example, $CuPc_{(s)}$; an electron from the cathode electronic conductor (e.g., a graphene grain or a carbon fiber of high electronic conductivity) enters the active material during the cell discharge period. After transfer of a lithium ion to a vacant active site on the solid cathode active material surface, it migrates, via configurational diffusion, to a most suitable site in the active material while in the active material, molecular electronic configuration readjustment also takes place. The quantum mechanical scale level adjustments of electrons occur in such a way that the occupation of a most suitable site by the intercalated lithium ion satisfies the required thermodynamic condition of minimum energy of the cathode active material system.

The configurational diffusion of a lithium ion in the channels of copper phthalocyanine, with the channel size range of the order of 3 – 5 Å, is imagined to involve the forces between the migrating lithium ion and the solid copper phthalocyanine atoms in its local environment. The lithium ion movement is facilitated and made possible from one location to the next by the thermally activated cooperative motion of the neighboring copper phthalocyanine atoms. Lithium ion migration in solid copper phthalocyanine, similar to atomic migration in solid materials, may be thought to resemble chemical changes; yet differs from them in that the bonds formed and broken between the migrating lithium ion and the copper phthalocyanine crystal atoms are usually not of a covalent nature. Under the steady-state isothermal and isobaric conditions, the lithium ion diffusion coefficient, D_{Li^+} , is here assumed to follow the empirical law of the type given below.

$$D_{Li^+} = D_{Li^+,0} \exp\left(-\frac{E_{A,d}}{RT}\right) \quad (1)$$

where R = universal gas law constant, T = temperature in K, $D_{Li^+,0}$ = pre-exponential factor, and $E_{A,d}$ = activation energy associated with the lithium ion diffusion in the copper phthalocyanine interstitial transport channels. At a given temperature, $D_{Li^+,0}$ and $E_{A,d}$ are most likely to be function of the resultant energy of the lithium ion interactions with the atoms of copper phthalocyanine in its vicinity and Li^+ – Li^+ interactions. It is very highly likely that $D_{Li^+,0}$ and $E_{A,d}$ would change when lithium ions are stored in the copper phthalocyanine sub-nanometer size channels during the cell discharge period. In fact, D_{Li^+} is most likely to decrease with an increase in g-moles of lithium ions stored per g-mole of copper phthalocyanine. This type of view is also observed [5] in connection with the host layered intercalation materials such as $FePO_{4(s)}$ and $CoO_{2(s)}$.

Since 2016, experimental and theoretical/modeling investigations have been performed to characterize the overall electrochemical behavior of $CuPc_{(s)}$ as the active material in a lithium-based cell: $Li_{(s)}$/electrolyte/$CuPc_{(s)}$ as the active material in the cell composite cathode. Modeling/theoretical efforts have been devoted to understanding the effects of the processes; such as the lithium ion transfer across the cell electrolyte-cathode active material ($CuPc_{(s)}$) interface and lithium ion diffusive transport in the sub-nanometer scale channels of $CuPc_{(s)}$; on the charge storage capacity of $CuPc_{(s)}$. In this rather relatively short presentation; the model formulation for the lithium ion charge transfer electrokinetics across the interface between the cell electrolyte and the cell cathode active material, $CuPc_{(s)}$, the Gaussian/QE/VASP computational/simulation results on lithium atom incorporation into a CuPc molecule, and the experimental cell voltage versus the charge storage capacity of the active material, $CuPc_{(s)}$, are provided in the following Sections 2 through 4.

2.0 Lithium Ion Charge Transfer Electrokinetics Across the Interface between the Cell Cathode Active Material, $CuPc_{(s)}$ and its Electrolyte

To explain the observed experimental data on the cell voltage versus charge storage capacity of a cathode active material, such as copper phthalocyanine ($CuPc_{(s)}$); understanding of the effects of the processes: lithium ion transport in the electrolyte, lithium ion transfer across the electrolyte-cathode active material, and the lithium ion diffusive transport in the sub-nanometer scale channels of the active material, $CuPc_{(s)}$, of a lithium-based cell: $Li_{(s)}$/electrolyte/$CuPc_{(s)}$ – cathode active material, is essential. An advanced level qualitative understanding of these processes must be translated into the formulation; which in turn is required for the analysis of electrochemical performance behavior of a given lithium-based cell/battery as well as for the design of such a system. The development of the rate law formulation for the lithium ion charge transfer across the electrolyte-$CuPc_{(s)}$ cathode active material given here is based on the generalized concept of vacant and occupied sites used in the solid-state surface chemical reactions [4].

The overall process of charge transfer at the contact point of the solid cathode active material, $CuPc_{(s)}$, electrolyte (in liquid or solid phase), and electronic conductor (such as carbon fibers, carbon nanotubes (CNTs), or graphene flakes); is represented by:

$$Li^{+} + S_{v} + e^{-} \underset{k_b^a}{\overset{k_f^c}{\rightleftarrows}} (Li^{+} \cdots e^{-}) - S_{occup} \tag{2}$$

where Li^+ = lithium ion in the electrolyte in contact with the cathode active material, copper phthalocyanine ($CuPc_{(s)}$), S_v = a vacant active site in the solid phase, $CuPc_{(s)}$, at the interface between the electrolyte and the cathode active material copper phthalocyanine, and $(Li^{+} \cdots e^{-}) - S_{occup}$ = active site occupied by a lithium ion. It is, here, assumed that the only reaction occurring at the three phase boundary contact point is the one given by Eq, (2). All other parasitic reactions, if any, involving lithium ions and the cathode electrode active material, $CuPc_{(s)}$, are either absent or assumed of negligible occurrence.

At the three phase contact point, the forward reaction rate, r_f^{echem}, per unit effective interfacial area of the cathode electrode active material is appropriately assumed to be proportional to the lithium ion molar concentration in the electrolyte, C_{Li^+}, as well as the vacant active site molar concentration, C_{S_V}, for the lithium ion intercalation into the electrode active material, The backward or reverse reaction rate, r_b^{echem}, per unit effective interfacial area is proportional to the molar concentration, $C_{(Li^{+} \cdots e^{-}) - S_{occup}}$, of the $CuPc_{(s)}$ active material sites occupied by not fully charged lithium ions in the active material (i.e., each lithium ion has charge less than (+1)). Note that here our focus is on the charge transfer process across the interface between the electrolyte and active material, $CuPc_{(s)}$, at the constant temperature and pressure conditions. The net rate of the electrochemical reaction, Eq. (2), in the forward, cathodic direction is given by

$$\begin{aligned} r_{net}^{echem} &= r_f^{echem} - r_b^{echem} \\ &= \left[k_f^c C_{Li^+} C_{S_V} \exp\left(-\frac{\alpha_c F}{RT} (\Phi_{cath-electrod} - \Phi_{ref-electrod}) \right) \right] \\ &\quad - \left[k_b^a C_{(Li^{+} \cdots e^{-}) - S_{occup}} \exp\left(\frac{\alpha_a F}{RT} (\Phi_{cath-electrod} - \Phi_{ref-electrod}) \right) \right] \end{aligned} \tag{3}$$

where r_{net}^{echem} = net cathodic direction reaction per unit interfacial area, [g-mole cm_s^{-2} s^{-1}], $(\Phi_{cath-electrod} - \Phi_{ref-electrod})$ = cathode electrode electric potential relative to a reference electrode such as: $Li_{(s)}/Li^+$ in 1 molar solution of the cell electrolyte; α_c = cathodic direction charge transfer coefficient = $n\beta$ = $(1)\beta$ = β, n = number of electrons involved in the electrochemical reaction process per occurrence of reaction, Eq. (2), in the forward or backward direction = 1 here, β (the symmetry factor, typically set as) = 1/2 ; α_a = anodic direction charge transfer coefficient = $n(1 - \beta)$ = $(1 - \beta)$ = ½ for β = ½, k_f^c, k_b^a = chemical cathodic and anodic direction rate coefficients, respectively, at the electrode temperature condition, F = Faraday constant = 96487

Coulomb of charge per g-mole electrons, R = universal constant = 8.314 J g-mol^{-1} K^{-1}. Net interfacial current density, i_s [A cm$_s^{-2}$], is given by

$$i_s = Fr_{net}^{echem} = F\left[k_f^c C_{Li^+} C_{S_V} \exp\left(-\frac{\alpha_c FE}{RT}\right) - k_b^a C_{(Li^+\cdots e^-)-S_{occup}} \exp\left(\frac{\alpha_a FE}{RT}\right)\right] \qquad (4)$$

where E = $(\Phi_{cath-electrod} - \Phi_{ref-electrod})$, [Volts].

At a cathode electrode active material state represented by X [g-mole of lithiated active sites per g-mole of the active material, $CuPc_{(s)}$]; the site balance is represented by:

$$(total\ active\ sites) = \left[(vacant\ active\ sites) + (lithiated\ sites)\right] \qquad (5)$$

Using the detailed information in Reference 3, Eq. (4) was transformed into Eq. (6).

$$i_d = \left[\frac{\sqrt{X(X_{\max} - X)}}{1+\alpha X}\right]\left[e^{-\zeta} - e^{\zeta}\right] \qquad (6)$$

where $\zeta = \frac{F\eta_s}{2RT}$ and $i_d = \left(\frac{i_s}{i_0}\right)$. Eq. (6) demonstrates the dependence of dimensionless current density, i_d, on X, ζ, and α [coefficient of volume expansion of active material].

Figure 1 shows the dimensionless current density over the range of X values at three values of ζ with an assumed X_{max} of 20 moles of active sites per mole of active material and a constant coefficient of volume expansion, α, of 0.005. The values of ζ are only for a discharging cell, where the surface overpotential, η_s, is less than zero. Increasing magnitude of ζ leads to increasing the value of the exponential term in Eq. (6). When temperature is held constant, this corresponds to increasing the magnitude of the surface overpotential. Because surface overpotential is the voltage deviation from the open circuit voltage, a greater magnitude discharge overpotential corresponds to a lower cell voltage. Therefore, under the conditions where surface charge transfer kinetics controls the overall cell discharge process rate, higher currents can be achieved by sacrificing electrical energy.

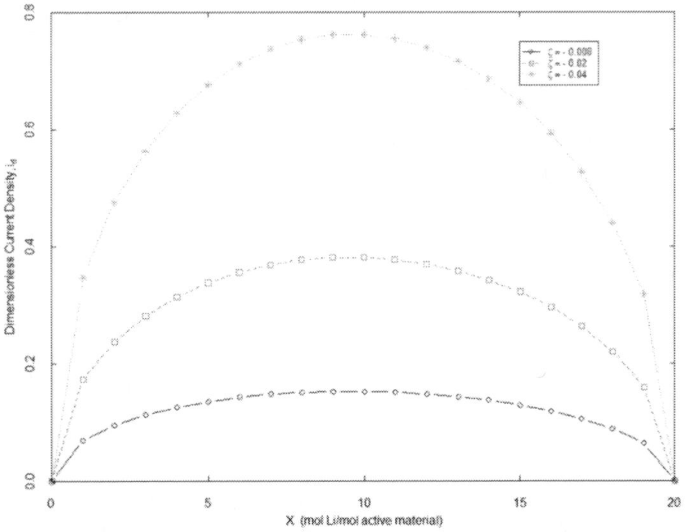

Figure 1. Dimensionless current density, i_d, vs X with X_{max} = 20 mol active sites per mol active material.

3.0 The Gaussian Computational Work on Copper Phthalocyanine ($CuPc_{(s)}$)

For solid copper phthalocyanine (CuPc) and lithium intercalation number into CuPc up to four lithium atoms, the theoretical Gibbs energy was determined using the Gaussian 16 software package. The Heyd-Scuseria-Ernzerhof functional, HSEH1PBE, with 6-31+G(d,p) basis set on C, H, N, Li and the basis set aug-cc-pvdz on Cu. The simulation also included the effect of the lithium ions in the diethoxyethane (DEE) solvent using the Polarizable Continuum Model (PCM). The results are given in Table 1 below. The calculated cell voltage ranges from 1.5 to 2.0 volts. These voltages are the open-circuit voltages and are close to what is observed experimentally at very low current densities such as C/100.

Table 1. Calculated cell voltage at 298.15K using Gaussian 16.

Species (lowest energy)	Calculated Cell Voltage (V)
Li_1CuPc(triplet)	1.632
Li_2CuPc(doublet)	1.962
Li_3CuPc(triplet)	1.611
Li_4CuPc(doublet)	1.505

The voltages in Table 1 correspond to optimized geometrical configurations. An example of an optimized configuration is shown below in Figure 4 for two lithium atoms intercalated into the copper phthalocyanine molecule which results in $CuLi_2Pc$. Note from the right side view, that there is some, but minimal, distortion of the normally planar molecular structure.

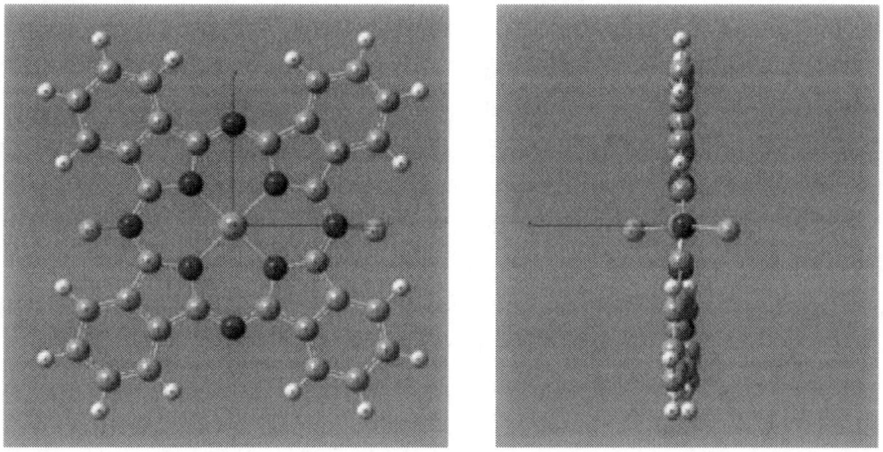

Figure 2. Optimized geometrical configuration of $CuLi_2Pc$ molecule using Gaussian 16 (top view and right-side view).

4.0 Experimental

Figure 3 shows the discharge of a lithium/CuPc coin cell that uses lithium foil as the anode, 1 molar lithium bis(fluorosulfonyl)imide (LiFSI) in dimethoxyethane (DME) as the electrolyte and Celgard 2300 as the separator, and 60 wt% CuPc (purity greater than 99%) as the cathode active material with 30 wt% carbon black as the electronic conductor and 10 wt% PVDF as the binder. The cell is first discharged at a C/10 rate to a 1.0 volt cut-off. The cell is then placed on open-circuit for two hours before performing a low discharge rate (C/96) to a 1.0 volt discharge cut-off to determine the remaining residual capacity. In this cell configuration at the C/10 rate, about 550 mAh/g of CuPc is observed with a total capacity, including residual capacity, is about 840 mAh/g of CuPc.

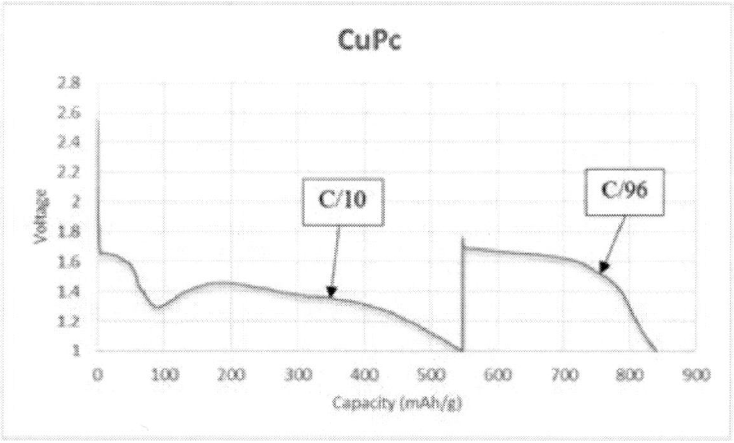

Figure 3. Coin cell discharge curve of 60 wt% CuPc with 30 wt% carbon black and 10 wt% PVDF.

Figure 4 shows the discharge of a hybrid lithium/CuPc-CF_x (x=1) coin cell that uses lithium foil as the anode, 1 molar LiFSI in dimethoxyethane (DME) as the electrolyte and Celgard as the separator, and 60 wt% CuPc-CF_x mixture as the cathode active material with 30 wt% electronic conductor being either carbon black or graphene, and 10 wt% PVDF as the binder. Note that the weight ratio of CF_x to CuPc, in this case, was 7:3. The total capacity of the hybrid cell using carbon black as the conductive material was almost 1500 mAh/g of active material which is about 1.75 times greater than the case of using CuPc alone as the active material. Using graphene as the conductive material in this type of hybrid cell results in a total capacity of almost 3800 mAh/g of active material; a factor of about 4.5 times greater than using CuPc alone.

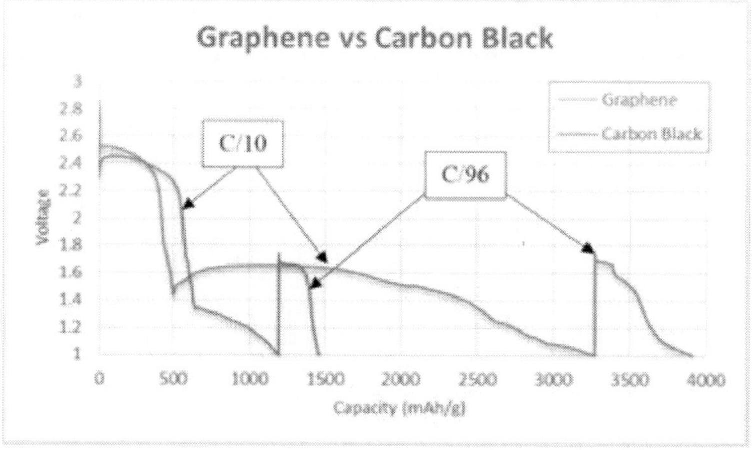

Figure 4. Comparison of coin cell discharge curves of graphene and carbon black as conductive additives, with a 7:3 weight ratio CFx to CuPc; 60 wt% total active material, 30 wt% conductive material, and 10 wt% PVDF.

5.0 Conclusions

The sub-model formulation for the lithium-ion charge transfer across the electrolyte-cathode active material for a lithium-based cell: $Li_{(s)}$/electrolyte/cathode active materials such as copper phthalocyanine has been presented for its incorporation into a complete cell model to predict the performance behavior of such a cell. The model based calculated data show the following: the maximum dimensionless discharge current density increases with an increase in the intercalated lithium content of the active material at a fixed dimensionless cathode polarization or overvoltage for a fixed set of cell temperature and pressure conditions; also, higher availability of active sites in the cathode active material leads to a higher dimensionless current density at a fixed lithium content of the cathode active material.

Gaussian 16 has been used to calculate the optimal geometrical configurations for lithium intercalation number into CuPc up to four lithium atoms per CuPc molecule and open-circuit voltages.

Capacity increases from 1.75 to 4.5 times the lithium metal/CuPc cell discharge capacity can be seen with the use of a hybrid lithium metal/CF_x-CuPc cell configuration.

Acknowledgement: We acknowledge the hard work and persistence provided by Max Tsao from the University of Dayton Research Institute.

6.0 REFERENCES

1. Sandhu, S.S, Fellner, J.P. (2015), J. Chem. Eng. Process. Technol., 6, 257.
2. Sandhu, S.S., Fellner, J.P., Tsao, M., Cashion, C.J. (2017), International Journal of Engineering Technology, Management and Applied Sciences (IJETAS), 5, 24.
3. Fellner, J.P, Scanlon, L.G., Tsao, M., Tsao, B., Sandhu, S.S., Lawson, J. (2017), AFOSRLRIR 16RQCOR301 Annual Report.
4. Froment, G.F., Bischoff, K.B., Wilde, J. De (2011), Chemical Reactor Analysis and Design (3rd Edition), John Wiley & Sons, Inc., pp. 64-75, 173
5. Ven, A.V.D., Bhattacharya, J., Belak, A.A. (2013), Accounts of Chemical Research, 46, 1216.

532214 Autocatalytic Reactions and Surface Diffusion Control Phase Separation in Li_XFePO_4

Abstract:

Many battery electrodes phase-separate upon lithium insertion and removal. Phase separation within electrodes arises when lithium ion/electron am bipolar pairs prefers to aggregate near other lithium ion/electron ambipolar pairs, often creating coexisting domains with different lithium stoichiometries within a single battery particle. Phase separation lead to chemo-mechanical stresses that may reduce cycleability. Li_XFePO_4 ($0<X<1$) is a model electrode that phase-separates into Li-rich and Li-poor domains under equilibrium. Under non-equilibrium conditions of lithium insertion and removal, we show the crucial role that surface reaction and surface diffusion play in phase separation kinetics, and how these parameters can be used to altogether suppress phase separation.

We developed synchrotron *operando* X-ray microscopy to track lithium insertion and migration within individual particles at ~50 nm spatial resolution and ~ 30 sec temporal resolution. We directly map the surface reaction kinetics as a function of the lithium stoichiometry as the battery particles charge and discharge. Our results show that lithium extraction (charge) is auto-catalytic while lithium insertion (discharge) is auto-inhibitory. The auto-catalytic charge amplifies the intrinsic tendency of Li_XFePO_4 to phase-separate, while auto-inhibitory discharge suppresses phase separation. When auto-inhibitory behavior is sufficiently strong at elevated rates of lithium insertion, separation can be completely suppressed and replaced with a solid solution. This shows how how surface reaction can be used to control a bulk material property.

For auto-catalytic reactions and for insufficiently auto-inhibitory reactions, phase separation can also be suppressed by reducing the ambipolar transport of lithium ion/electron pairs. Using both experiment and phase-field modeling, we show that the effective Damkohler number in this solid electrode is a crucial parameter at determining whether or not a particle phase-separates. When lithium diffusion is faster than lithium insertion, then the particles phase-separate. When lithium reaction is faster than diffusion, then the solid solution is kinetically stabilized because the particle does not have time to phase-separate before it finishes (de)lithiation. Experimental results show that this ambipolar transport is dominated by lithium surface diffusion at the particle/electrolyte interface. Thus, reducing surface diffusion and ambipolar transport provides another route towards controlling phase separation in Li_XFePO_4.

535165 Optimal Thermal Management of a High-Temperature Sodium Sulphur Battery

Abstract:

Optimal thermal management of a high-temperature sodium sulphur battery

Sai Pushpitha Vudata, Debangsu Bhattacharyya, Richard Turton

Department of Chemical and Biomedical Engineering, West Virginia University, Morgantown, WV 26506, USA

The sodium sulfur battery is an advanced secondary battery with high potential for grid-level storage due to their high energy density, low cost of the reactants, and high open-circuit voltage. However, as the current density is high and the normal operating temperature of the battery is high (about 300°C), effective thermal management is required to prevent thermal runaway. This work proposes a hybrid cooling strategyâ€" a combination of active and passive methods- that not only results in effective thermal management but also utilization of the heat for better energy efficiency. The cooling strategies will be implemented by an optimal method based on a control variable.

To study the transient response of sodium-sulfur batteries under high current density operation and to develop efficient optimal thermal management strategies, a detailed, thermo-electrochemical dynamic model of a sodium-sulfur cell is developed. However, the cell model is computationally intractable for simulating the large number of cells in the battery. Various strategies such as coordinate transformation, orthogonal collocation, and model reformulation (Northrop et al., 2011) are implemented to obtain a reduced order model that solves significantly faster than the full, high-dimensional model but provides an accurate estimate of the key variables.

Under rapid charging/discharging, especially under discharging conditions, it can lead to significant excursion in the battery temperature when the current density is high. Three thermal management strategies are considered in this work: active cooling, passive cooling, and hybrid cooling. The active cooling strategy uses air as the cooling medium whereas, the passive cooling strategy uses a phase change material (PCM) to serve as the thermal storage offsetting the transients. The passive cooling strategy provides faster heat dissipation than the active cooling strategy especially during high pulse power discharges while maintaining sufficiently uniform cell temperature. However, it was observed that the passive cooling strategy can lead to high heat accumulation in the PCM especially under very high current density and discharging conditions. This can eventually lead to thermal runaway. A hybrid cooling strategy is developed that can not only maintain the cell temperature near the optimum under rapid cycling operation at high current density, but can effectively utilize the heat improving the overall efficiency of the battery system.

Reference

- Northrop PWC, Ramadesigan V, De S, and Subramanian VR, â€œCoordinate Transformation, Orthogonal Collocation, Model Reformulation and Simulation of Electrochemical-Thermal Behavior of Lithium-Ion Battery Stacksâ€, 158, A1461-A1477, Journal of the Electrochemical Society, 2011

534973 Stochastic Statistical Models of Vehicle-to-Grid Economics for Predicting Impact of Policy and Renewables Portfolio

Abstract:

The continued growth of renewable sources of energy lead to grid-level intermittent supply and demand mismatch. There are a number of research and engineering solutions that address this problem including grid-level energy storage and demand response via a smart grid infrastructure. Vehicle-to-grid (V2G) is a promising approach because it uses the existing resource of electric vehicle batteries as the energy storage medium. Electric vehicles charge at night while power is cheap, commute to work, and discharge when demand is high yielding the electric vehicle owner a profit. This simple sounding process has a number of complicating economic issues and in this talk we discuss them. For example, charge/discharge efficiency, battery degradation, and location-based marginal pricing of electricity. Our preliminary conclusion is V2G is not beneficial to an electric vehicle owner with current technology and electricity pricing, even with public policy tools. In our model, a stochastic economic analysis of V2G which takes into account randomness in driving patterns and work patterns of EV users is presented. We use battery degradation models to study the impact of vehicle battery degradation on this process. This analysis is done for multiple cities of the United States and location based results are presented. We also apply an economic analysis at the utility level to see how V2G affects load and what issues can arise. For example, as solar makes up more of the electricity generation mix, the highest price of electricity may be at night when generation is minimal. V2G can improve the peak and demand mismatch in such cases.

533889 Electrochemical and Thermal Modeling of Capacity Fade in Lithium Ion Batteries for Prognosis – a Reaction Kinetic Approach

Abstract:

Electrochemical and thermal modeling of capacity fade in Lithium ion batteries for prognosis – A reaction kinetic approach

Parth Shah[1], Sathish Swaminathan[1], Resmi Suresh[1], Raghunathan Rengaswamy[1]

[1]Department of Chemical Engineering, IIT Madras,

Chennai, India.

Email address: raghur@iitm.ac.in

Lithium ion batteries have found widespread use in automotive industry owing to their high energy densities, longer life and reducing costs. However they are subject to unwarranted capacity loss, which is a major cause of concern for the automotive industry. Conventional battery lifetime analysis involves testing the batteries for thousands of hours using expensive and sophisticated instrumentation[1]. These tests are performed under ideal laboratory setup and do not capture variations in operating condition. However lithium ion batteries are highly sensitive to changes in operating conditions, which further affects battery life. This work extends on [2] in developing a comprehensive model to capture the capacity fade in lithium ion batteries across varying operating conditions. The model incorporates both, the electrochemical and thermal aspects. A reaction kinetics based approach has been adopted which allows for inclusion of side reactions in the capacity fade mechanism. Additional capacity fade mechanisms and failure modes could also be incorporated in the given framework allowing for quick and efficient prognosis.

References:

[1] Burns JC, Kassam A, Sinha NN, Downie LE, Solnickova L, Way BM, et al. Predicting and Extending the Lifetime of Li-Ion Batteries. J Electrochem Soc 2013;160:A1451–6. doi:10.1149/2.060309jes.

[2] Suresh, R., Rengaswamy R. Modeling Failure Modes in Li-Ion Battery. 2015 AIChE Annu. Meet., Salt Lake City, UT, USA: 2015, p. 287h.

533737 The Effect of Loading, Test Parameters, and Oxides on Electrolyzer-Catalyst Durability

Abstract:

In commercial electrolysis, the cost of electricity input drives the cost of hydrogen production. Therefore, electrolysis is typically run at high catalyst loading and constant power input over long periods of time. Lowering water-splitting hydrogen production costs to a level comparable to steam methane reformation requires: coupling electrolysis with low-cost power input (wind, solar) to reduce feedstock costs; and dropping catalyst loading to reduce the capital cost at lower capacity.[1,2] While minimal durability loss is seen in commercial electrolyzers, catalyst losses can be masked by high loading (several mg cm ^{2}). At low loading, however, these losses become more apparent.[3]

In this study, electrolyzer durability was evaluated at low iridium-anode loading (0.1 0.5 mg cm ^{2}) and with different power inputs (potentials, intermittency). Higher loading tended to delay the onset of durability losses; in contrast, a loading of 0.1 mg cm_{elec} ^{2} resulted in incremental but immediate loss when exposed to high potential. Increasing the upper potential limit appeared to increase iridium dissolution and migration and resulted in higher performance losses. This trend was expected and iridium dissolution is anticipated to be a primary factor in electrolyzer loss at low loading.[4] Introducing cycling (square/triangle wave), however, significantly increased the rate of performance decay and was less expected from half-cell tests and even though less time was spent at elevated potential.[3] Changing the rate of potential increase (saw tooth profiles) confirmed that rapid input increases accelerated loss and may be due to localized potential spikes occurring within the catalyst layer. A variety of system control-based mitigation strategies have been evaluated for lessening durability losses when handling intermittent power sources.

Several iridium catalyst types (oxides, surface areas) have been tested in half- and single-cells. Catalyst development efforts often focus on metallic- or hydroxide-based iridium structures, due to higher activity in ex-situ tests. These performance advantages, however, largely disappear in single-cell tests and may be due to surface/near-surface oxidation during conditioning protocols. Metallic/hydroxide durability losses further tend to be larger in both half- and single-cells and may be related to the kinetics of iridium metal/hydroxide/oxide dissolution.

In single-cell testing, we have focused on how loading, test parameters, and catalyst type affect proton exchange membrane-based electrolyzer durability. These tests have significant implications on lowering the cost of electrolysis-based hydrogen production and on coupling electrolysis with renewable power inputs.

[1] H2 at Scale: Deeply Decarbonizing our Energy System. Presented at Annual Merit Review, U.S. Department of Energy; Washington, DC, June 6–10, 2016. https://www.hydrogen.energy.gov/pdfs/review16/2016_amr_h2_at_scale.pdf.

[2] Denholm, P.; O'Connell, M.; Brinkman, G.; Jorgenson, J. Overgeneration from Solar Energy in California: A Field Guide to the Duck Chart; Vol. NREL/TP-6A20-65023; National Renewable Energy Laboratory: Golden, CO, 2015. Available at the following: http://www.nrel.gov/docs/fy16osti/65023.pdf.

[3] Alia, S. M.; Rasimick, B.; Ngo, C.; Neyerlin, K. C.; Kocha, S. S.; Pylypenko, S.; Xu, H.; Pivovar, B. S. *J. Electrochem. Soc.* **2016**, 163, F3105–F3112.

[4] Cherevko, S.; Geiger, S.; Kasian, O.; Kulyk, N.; Grote, J. P.; Savan, A.; Shrestha, B. R.; Merzlikin, S.; Breitbach, B.; Ludwig, A.; Mayrhofer, K. J. J. *Cat. Today* **2016**, 262, 170.

537181 Thermodynamic Modeling of Electric Double Layer in Capacitive Deionization Cell Electrodes with Condensation Theory

Abstract:

Capacitive deionization (CDI) is a novel technology to purify brine water of low salinity with electric cell units. Ionic components are attracted to the electrodes when the cell units are charged and released when disconnected. The modified Donnan (mD) model has been used to model the electrical double layer (EDL) and CDI units by Biesheuvel et al. The mD model describes the electric double layer as a homogeneous solution of counterions and coions. Furthermore, thermodynamic activity coefficients of mobile ions in the electrodes have not been considered. [1, 2]

In this work, we propose the electric double layer consist of a Stern layer and a diffuse layer. The counterions in the Stern layer are considered tightly attached (or condensed) to the surface of the electrode, while the diffuse layer contains free counterions and small amount of coions which are in Donnan equilibrium with the bulk electrolyte solution. The scenario is analogous to that of aqueous polyelectrolytes, where counterion condensation happens when the charge density of the polyelectrolyte is higher than a critical point above which the counterions condense to the surface of the polyion. As a limiting law, Manningâ€™s counterion condensation theory has been shown to describe the condensation phenomenon and calculate the activity coefficients of mobile ions in polyelectrolyte solutions. [3] Assuming the Stern (condensed) layer is a plate capacitor and equality of the chemical potentials of solutes in the bulk solution and in the diffuse layer, the amount of condensed counterions, Stern and Donnan potential drops, salt adsorption capacity and electrode surface charge density can be calculated with the condensation model. The nonideality of mobile ions can also be captured with Manningâ€™s counterion condensation theory. Experimental measurements of CDI performance for aqueous NaCl and $CaCl_2$ solutions of various concentrations under different cell voltages are conducted. The salt adsorption data can be well explained with our model.

Reference

1. Biesheuvel, P. M., et al. "Attractive forces in microporous carbon electrodes for capacitive deionization."*Journal of solid state electrochemistry* 5 (2014): 1365-1376.
2. Biesheuvel, P. M., H. V. M. Hamelers, and M. E. Suss. "Theory of water desalination by porous electrodes with immobile chemical charge."*Colloids and Interface Science Communications* 9 (2015): 1-5.
3. Manning, Gerald S. "Limiting laws and counterion condensation in polyelectrolyte solutions I. Colligative properties."The journal of chemical Physics 3 (1969): 924-933.

545759 Development of Devices and Selective Catalysts for the Solar-Driven Reduction of CO_2 to fuels

Abstract:

The electrochemical and photoelectrochemical reduction of CO_2 presents an attractive route towards fuel formation from renewable energy. However, the dearth of selective, active and inexpensive catalyst materials remains a key challenge on the way to renewable production of hydrocarbon fuels. Here, we address these challenges through the design of novel devices and selective earth-abundant catalysts for the electrochemical[1–3] and solar-driven[1,4–6] reduction of CO_2. In addition, we provide insight into the origin of selectivity on electrocatalysts based on copper, paving the way towards more rational design of electrochemical fuel forming devices.

We pioneer the application of Atomic Layer Deposition (ALD) to direct the catalytic selectivity of copper catalysts. Copper-based catalysts derived from the reduction of CuO are known to reduce CO_2 but simultaneously exhibit poor selectivity. ALD coating of CuO with thin layers of SnO_2 allows for formation of CO with high selectivity even at low overpotentials, comparing favorably to conventional gold catalysts, known to selectively convert CO_2 to CO. We use kinetic studies, high resolution transmission electron microscopy and gas chemisorption analysis to explain the observed selectivity changes on the basis of decreased binding strengths for both H and CO intermediates on the Sn-containing catalyst surface.

We subsequently combine our new catalyst with an earth abundant oxide anode to effect overall splitting of CO_2 to CO and O_2. Driven by a 3-junction solar cell, this device achieved long-term solar splitting of CO_2 into CO and O_2 at an efficiency of 13.4 %, constituting the current world record.[1]

Going beyond the production of CO, we investigate the further reduction of CO to hydrocarbons. Insights into the factors controlling the branching between different hydrocarbon species and hydrogen as a side product are scarce, precluding any rational design of selective catalysts. To remedy these shortcomings, we devise a platform allowing for the systematic study of the relevant reaction parameters. Exploiting non-aqueous electrolytes, we isolate activation-controlled kinetic data in electrochemical driving force, proton availability, solvent dielectric and reagent concentration. Our observations provide unique insight into the mechanism of hydrocarbon production on copper electrodes, demonstrating that the competition between H and CO for surface sites plays an important role in governing the product selectivity between methane, ethylene and hydrogen. These results establish a new paradigm in controlling hydrocarbon product selectivity in electrochemical fuel formation.[2]

References

[1] M. Schreier, F. Héroguel, L. Steier, S. Ahmad, J. S. Luterbacher, M. T. Mayer, J. Luo, M. Grätzel, *Nat. Energy* **2017**, *2*, 17087.

[2] M. Schreier, Y. Yoon, M. N. Jackson, Y. Surendranath, *Angew. Chem. Int. Ed.* **2018**, DOI 10.1002/anie.201806051.

[3] G. P. Lau, M. Schreier, D. Vasilyev, R. Scopelliti, M. Grätzel, P. J. Dyson, *J. Am. Chem. Soc.* **2016**, *138*, 7820–7823.

[4] M. Schreier, P. Gao, M. T. Mayer, J. Luo, T. Moehl, M. K. Nazeeruddin, S. D. Tilley, M. Grätzel, *Energy Env. Sci* **2015**, *8*, 855–861.

[5] M. Schreier, L. Curvat, F. Giordano, L. Steier, A. Abate, S. M. Zakeeruddin, J. Luo, M. T. Mayer, M. Grätzel, *Nat. Commun.* **2015**, *6*, 7326.

[6] M. Schreier, J. Luo, P. Gao, T. Moehl, M. T. Mayer, M. Grätzel, *J. Am. Chem. Soc.* **2016**, *138*, 1938–1946.

530409 *Ab-Initio* Investigation of Dimethyl Disulfide As an Additive for Lithium-Sulfur Batteries

Abstract:

Energy storage technology is an important field of study with strong economic and environmental driving forces. Batteries are ubiquitous and critical for much of modern technology like cell phones and electric vehicles. The current leader in rechargeable battery technology is the lithium ion (Li) battery. However, the Li battery is close to the theoretical limit of its energy storage but our technology requires more capacity. One promising battery system is the Lithium-Sulfur (LiS) battery which can surpass the Li battery by about 3 times the energy density.

Unfortunately, the LiS battery faces several challenging problems before it can be utilized commercially. One problem is known as the polysulfide shuttle reaction. This reaction is the consequence of the solubility of sulfur reaction products (polysulfides) in the electrolyte. The polysulfide shuttle causes active material loss from the cathode and passivates both electrodes. Donghai Wang's group at Penn State University recently published a new strategy to counteract this effect: use a different electrolyte. Their new electrolyte consisted of 50% DME and 50% dimethyl disulfide (DMDS) which showed increased battery performance. Wang et. al's experiment postulated several potential reasons for why they saw increased performance but were not able to determine exactly why.

We used density functional theory (DFT) to explore the theory and fundamentals behind DMDS's effect on battery performance to complement Wang et al's experimental work. Computational chemistry calculations were completed with Gaussian09 to determine electron affinities and how the reduction mechanism is modified from reducing polysulfides to reducing dimethyl polysulfides. . We also investigated the passivation and solubility of new reduction products on the electrodes with the DMDS modified reduction pathway by solid state calculations with VASP. Based on the DOS and the mobility of charge carriers like hole polarons and charged defects we show that even with a modified reaction, the electrode will still be passivated. Regarding the reversibility of the reactions, the new electrode precipitate was determined to be more soluble than Li_2S which indicates that the precipitate is more reversible. Together these studies provide insight into how the LiS battery performance can be improved with the addition of DMDS.

531204 Atomic Layer Deposition of Protective Coatings on $LiMn_2O_4$ Cathodes

Abstract:

Spinel $LiMn_2O_4$ (LMO) is a low cost, non-toxic, and thermally stable lithium ion battery electrode with reasonable energy and power densities. Despite these advantages, however, its cycling performance is limited, due in large part to Mn dissolution from the electrode surface.[1] Mn dissolution is proposed to proceed following charge disproportionation between two Mn^{3+} cations, resulting in Mn^{4+} (insoluble) and Mn^{2+} (soluble) ion formation.[2] In turn, protective surface coatings have emerged as a promising strategy to mitigate Mn dissolution. Coatings have been proposed to (1) increase the concentration of surface Mn^{4+} ions to inhibit Mn^{3+} disproportionation, and (2) provide a physical barrier between Mn ions and the electrolyte.[3] Despite the enhanced performance of coated electrodes, there remains a lack of fundamental understanding regarding the registry of protective films to the electrode surface, as well as the role of the electrode/film interfacial structure towards electrochemical performance.

In this work, first principles density functional theory (DFT) calculations are used to evaluate the formation of Al_2O_3 coatings by atomic layer deposition (ALD), a self-limiting vacuum technique for thin film growth. Al_2O_3 film growth is considered through the prototypical reaction sequence of alternating trimethylaluminum (TMA) and H_2O half-reactions. DFT calculations, along with *in situ* experiments (FTIR, QMS, QCM), demonstrate that the TMA precursor undergoes significant decomposition in early ALD cycles, leading to site blocking sub-monolayer coverages of Al_2O_3. Through electrochemical cycling, LMO electrodes with sub-monolayer coatings show enhanced capacity in comparison to fully conformal films formed upon further ALD. Through DFT studies of reaction pathways on thermodynamically stable[4] (001), (111) terraces, in addition to (511) step models, we outline the structure sensitivity of film growth during the early stages of ALD. The findings show that film nucleation is most facile at the more reducible surface facets, that given their near-surface electronic structure, are also most likely to undergo disproportionation and dissolution. These trends help provide an explanation for enhanced cycling performance with only 1-2 ALD cycles, wherein defects on the LMO surface may be selectively passivated. We will address how the trends outlined for the TMA/H_2O system can be extended to control the onset of film growth based on precursor selection.

This research was supported as part of the Center for Electrochemical Energy Science, an Energy Frontier Research Center funded by the U.S. Department of Energy, Office of Science, Office of Basic Energy Sciences, under Contract No. DE-AC02-06CH11357.

References

1. C. Zhan, T. Wu, J. Lu, and K. Amine, *Energy Environ. Sci.*, **11**, 243â€"257 (2018).
2. J. C. Hunter, *J. Solid State Chem.*, **39**, 142â€"147 (1981).
3. L. Jaber-Ansari et al., *Adv. Energy Mater.*, **5**, 1500646 (2015).
4. R. E. Warburton, H. Iddir, L. A. Curtiss, and J. Greeley, *ACS Appl. Mater. Interfaces*, **8**, 11108â€"11121 (2016).

525559 Silicon Li-Ion Anode Materials Via Spray Drying and Magnesiothemic Reduction

Abstract:

We report an effective and scalable method to produce silicon (Si) and silicon-carbon (Si-C) composite with the combination of spray drying, magnesiothermic reduction and chemical vapor deposition. Commercial silica water suspension (LUDOX HS-40) is used as the Si precursor. Diluted silica (SiO_2) suspension is spray-dried with an aerosol spray dryer. Subsequent magnesiothermic reduction of the spray-dried SiO_2, with sodium chloride as the heat scavenger, is performed to convert SiO_2 to Si with retention of the secondary composite structure. After coating a thin layer of carbon via chemical vapor deposition, the Si-C composites as Li-ion battery anode materials show excellent cycle stability of 1400 mAh/g for 100 cycles.

534257 The Influence of Electrode Microstructure on the Performance of Non-Aqueous Redox Flow Batteries

Abstract:

Redox flow batteries (RFBs) are promising rechargeable electrochemical devices for grid-scale energy storage but further cost reductions are needed for ubiquitous adoption of this technology [1]. While current state-of-the-art RFB systems employ aqueous chemistries, transitioning to nonaqueous electrolytes offer a new pathway to reduced cost through increased energy density. While research efforts have primarily focused on molecular discovery [2], there has been significantly less attention paid to the development of other critical system components. Of particular importance are the porous electrodes used in the electrochemical stack. Todayâ€™s electrodes largely draw from the fuel cell material set, but, within a RFB, the porous electrode must perform a number of different roles including serving as active surfaces for electrochemical reactions, facilitating uniform liquid electrolyte distribution, and supporting low pressure drops [3]. Thus, before advanced electrodes can be developed for RFB applications, performance-limiting factors for the present materials set must be quantified. However, unambiguous analysis is challenging in an operating RFB due to the complex coupling of transport and reactions which varies as a function of state of charge during cell cycling. Deconvoluting the role of electrode properties on flow battery performance requires the development of diagnostic techniques that enable electrode characterization under well-controlled but application-relevant conditions.

To this end, we systematically compare the *operando* performance of RFBs containing selected carbon paper, felt, and cloth electrodes using the single-electrolyte cell configuration [4] and a model organic redox couple (TEMPO/$TEMPO^{+}$) [5]. Using polarization and electrochemical impedance spectroscopy, we quantify the impact of electrode microstructure on battery performance. We find that, depending on the electrode choice and flow conditions, current densities as high as 450 mA cm^{-2} can be achieved at an overpotential of 0.3 V with a cell area specific resistance as low as 0.7 Î© cm^{2}. This result suggests that, through appropriate cell engineering and materials selection, high power performance may be realized in nonaqueous flow batteries. Finally, a simple convection-diffusion mass transport model is developed to explain the observed experimental behavior and provide guidance for the bottom-up design of next generation RFB electrodes.

Acknowledgements

We gratefully acknowledge the financial support of the Swiss National Science Foundation (P2EZP2_172183) and the Joint Center for Energy Storage Research (JCESR), an Energy Innovation Hub funded by the United States Department of Energy.

References

1. Darling et al., *Energy Environ. Sci.*, 7, 3459 (2014).
2. A. Kowalski et al., *Curr. Op. Chem. En.*, 13, 45 (2016).
3. Weber et al., *J. Appl. Electrochem.*, 41(10), 1137 (2011).
4. R. Darling et al., *J. Electrochem. Soc.*, 161, A1381 (2014).
5. D. Milshtein et al., *J. Power Sources*, 327, 151 (2016).

535401 Method for Rapid Evaluation of Graphite Felt Electrode Stability in Vanadium Redox Flow Battery

Abstract:

Vanadium redox flow battery (VRFB) is potential electrochemical energy storage solution for residential batteries and grid stabilization. The long durability originating mainly from all-vanadium chemistry, represents the main advantage of the technology. The battery stack is constructed from several components, where membrane and graphite felt electrodes are largely responsible for its efficiency and durability.

In our contribution, we present the developed method for accelerated studies of degradation processes of felt electrodes and quantification of intensity of side reactions under defined conditions (state-of-charge, current density and temperature). Full-cell and half-cell characterization is used to evaluate the electrochemical changes of positive and negative electrode, separately. The method is applied to evaluate the effect of the battery operating conditions on the performance durability of various graphite felt materials in vanadium redox flow battery. Subsequent post mortem analysis of the felts in VRFB by relevant physico-chemical characterization techniques relates the performance decay with the changes in felt properties. We observed significant differences in the stability among individual felt materials and their treatment.

528204 Designing Electrocatalysts for Vanadium-Based Flow Batteries for Renewable Energy Storage

Abstract:

The world's growing population and rising standard of living will continue to increase energy demand. To satisfy energy demand, fossil fuels are burned at an incredible rate, emitting 35 billion tons of CO_2 in 2017. There are technologies capable of satisfying energy demand without CO_2 emissions, such as solar or wind, but they are intermittent in energy generation, thus their energy must be stored to match demand.

Vanadium-based redox flow batteries (VRFBs) are a promising technology to store renewable electricity. VRFBs (e.g., all vanadium, V-Ce, V-Br) have been tested on benchtop and at pilot scales. The main challenge for the widespread adoption of VRFBs is cost, which can be reduced by increasing the current densities without decreasing energy storage efficiency. Electrocatalysts have been used to accelerate the V^{2+}/V^{3+} redox reaction in VRFBs (shown in Equation 1), but a mechanistic understanding of the V^{2+}/V^{3+} redox reaction is still unclear.

V^{2+} V^{3+} + e^- with $E^0(V^{2+}/V^{3+})$ = -0.255 V vs. SHE

One possible mechanism on carbon, metal oxides, and metals involves the exchange of the vanadium ion with H (present as OH on carbon or metal oxides and metal-H on metals), where this exchange reaction acts as an intermediate step in the V^{2+}/V^{3+} reduction or oxidation reaction.

We will discuss experimental measurements of V^{2+}/V^{3+} redox activity on different electrocatalysts (e.g., glassy carbon, Au, Bi) and compare the results to density functional theory calculations of hydrogen adsorption. Our hypothesis is that the strength of the bound hydrogen should correlate with the electrocatalyst reaction rate, if H_{ads} plays a role in the V^{2+}/V^{3+} reaction. For metals, the energy of the adsorbed hydrogen intermediate is well-known to correlate with the hydrogen evolution activity (an undesired side-reaction in vanadium-based RFBs),[1,2] so we will use this to eliminate certain metals from consideration. Of the metals reported currently to be active for the V^{2+}/V^{3+} redox reaction (i.e., Bi,[3–7] Sb[8], Cu,[9] Sn[10]) all bind hydrogen very weakly according to our DFT calculations. We will discuss if other 'descriptors' (such as vanadium ion adsorption strength) are required to understand the V^{2+}/V^{3+} reaction rates on metal surfaces.

References

(1) Nørskov, J. K.; Bligaard, T.; Logadottir, A.; Kitchin, J. R.; Chen, J. G.; Pandelov, S.; Stimming, U. Trends in the Exchange Current for Hydrogen Evolution. *J. Electrochem. Soc.* **2005**, *152* (3), J23.

(2) Parsons, R. The Rate of Electrolytic Hydrogen Evolution and the Heat of Adsorption of Hydrogen. *Trans. Faraday Soc.* **1957**, *54* (0), 1053–1063.

(3) Yang, X.; Liu, T.; Xu, C.; Zhang, H. H.; Li, X.; Zhang, H. H. The Catalytic Effect of Bismuth for VO2+/VO2+ and V3+/V2+ Redox Couples in Vanadium Flow Batteries. *J. Energy Chem.* **2017**, *26* (1), 1–7.

(4) Suárez, D. J.; González, Z.; Blanco, C.; Granda, M.; Menéndez, R.; Santamaría, R. Graphite Felt Modified with Bismuth Nanoparticles as Negative Electrode in a Vanadium Redox Flow Battery. *ChemSusChem* **2014**, *7* (3), 914–918.

(5) Park, M.; Ryu, J.; Cho, J. Nanostructured Electrocatalysts for All-Vanadium Redox Flow Batteries. *Chem. - An Asian J.* **2015**, *10* (10), 2096–2110.

(6) Flox, C.; Murcia-López, S.; Carretero, N. M.; Ros, C.; Morante, J. R.; Andreu, T. Role of Bismuth in the Electrokinetics of Silicon Photocathodes for Solar Rechargeable Vanadium Redox Flow Batteries. *ChemSusChem* **2017**, *8028*, 125–129.

(7) Liu, T.; Li, X.; Nie, H.; Xu, C.; Zhang, H. Investigation on the Effect of Catalyst on the Electrochemical Performance of Carbon Felt and Graphite Felt for Vanadium Flow Batteries. *J. Power Sources* **2015**, *286*, 73–81.

(8) Shen, J.; Liu, S.; He, Z.; Shi, L. Influence of Antimony Ions in Negative Electrolyte on the Electrochemical Performance of Vanadium Redox Flow Batteries. *Electrochim. Acta* **2015**, *151*, 297–305.

(9) Wei, L.; Zhao, T. S.; Zeng, L.; Zhou, X. L.; Zeng, Y. K. Copper Nanoparticle-Deposited Graphite Felt Electrodes for All Vanadium Redox Flow Batteries. *Appl. Energy* **2016**, *180*, 386–391.

(10) Mehboob, S.; Mehmood, A.; Lee, J.-Y.; Shin, H.-J.; Hwang, J.; Abbas, S.; Ha, H. Y. Excellent Electrocatalytic Effects of Tin through: In Situ Electrodeposition on the Performance of All-Vanadium Redox Flow Batteries. *J. Mater. Chem. A* **2017**, *5* (33), 17388–17400.

534409 Increasing the Charge Storage Capacity of Phenothiazine-Based Electrolytes for Nonaqueous Redox Flow Batteries

Abstract:

Stationary energy storage systems are anticipated to play an important role in facilitating the integration of intermittent, renewable energy sources and in improving the efficiency, reliability, and resiliency of the existing fossil fuel based infrastructure. Redox flow batteries (RFBs) are rechargeable electrochemical devices that are well suited for grid storage due to decoupled power and energy scaling to meet specific installation needs, long operating lifetimes (~10 years), and simplified manufacturing[1]. Despite this promise, further cost reductions are needed for widespread adoption motivating research and development into new redox chemistries, electrolyte formulations, and reactor designs.[2, 3] The combination of organic redox active materials and nonaqueous electrolytes is an emerging concept that aims to lower system costs through increased energy density.[4] A key challenge is the design and implementation of stable, soluble, and scalable redox couples that enable high capacity redox electrolytes.

Here, using *N*-ethylphenothiazine and its derivatives as a learning platform, we systematically investigate the impact of substituent group addition on the molecular properties to develop structure-function relations that may enable deterministic multi-property optimization. Specifically, we couple molecular engineering and electrochemical analysis to increase the equivalent charge concentration of phenothiazine-containing electrolytes by enhancing the solubility[5] and intrinsic storage capacity[6]. We find that through careful selection and positioning of substituent groups, significant performance improvements may be realized. The lessons learned here are portable and can be used to curate design campaigns for other redox couples, hence accelerating the development of new RFB materials.

1. A. Z. Weber et al., J. Appl. Electrochem., 41, 1137–1164 (2011).
2. R. M. Darling, K. G. Gallagher, J. A. Kowalski, S. Ha, and F. R. Brushett, Energy Env. Sci, 7, 3459–3477 (2014).
3. L. Su, J. A. Kowalski, K. J. Carroll, and F. R. Brushett, in Rechargeable Batteries, Green Energy and Technology. Z. Zhang and S. S. Zhang, Editors, p. 673–712, Springer International Publishing (2015).
4. J. A. Kowalski, L. Su, J. D. Milshtein, and F. R. Brushett, Curr. Opin. Chem. Eng., 13, 45–52 (2016).
5. J. D. Milshtein et al., Energy Environ. Sci., 9, 3531–3543 (2016).
6. J. A. Kowalski et al., J. Mater. Chem. A, 5, 24371–24379 (2017).

537024 Investigating Cyclability of Aqueous Redox Flow Battery Electroactive Species at High Concentrations

Abstract:

Structure-function relationships for electroactive species have been useful in the development of redox flow batteries for grid level storage, particularly in terms of energy density. One of the remaining challenges is systematically determining how molecular structure and electrolyte composition impact stability. In this paper, we describe our efforts to understand the cyclability of various commercially available redox active molecules as a function of their concentration in aqueous media. The cyclability is assessed using cyclic bulk electrolysis and the decay products were characterized using cyclic voltammetry, UV-Vis, NMR and Raman spectroscopy. Quinones have emerged as very promising electroactive species for aqueous redox flow batteries. For anthraquinone 2-monosulfate (AQDS-2), a well-studied quinone analog, we found that at low concentrations the capacity fade is slow. In particular at 8% [50mM] and 25% [160mM] of max solubility, the decay rates are 0.005% and 0.024% per cycle, respectively. At 50% of the maximum solubility [320mM], the decay rate increased significantly to 0.5% of total material per cycle. These results are consistent with recently published reports which suggest the degradation is due to dimerization or higher order polymerization reactions. Flavins such as riboflavin-5'-monophosphate, and metal coordination complexes including ferrocenium and cobalt terpyridine derivatives displayed better capacity retention. For example, the riboflavin at 50% [50mM] of its solubility limit exhibited a decay rate of 0.014% per cycle in simple basic media. The addition of nicotinamide, a hygroscopic agent, improved the performance. Spectroscopic analysis of the spent solutions provides some insight regarding the degradation mechanisms. These and other results will be presented in the paper.

534534 Systems Approaches to Predict the Aqueous Solubility of Quinone Molecules for Flow Battery Applications

Abstract:

Flow batteries are electrochemical devices in which electrical energy is stored in the form of chemical energy in its electrolytes. During operation, the electrolytes are circulated through the flow battery cell and the stored chemical energy is converted to electrical energy through an electrochemical reaction. The major disadvantage of flow batteries over other storage devices is their low energy density. Energy densities of existing electrolytes can be improved by increasing their solubility in solvents. Identifying new electrolyte chemistries with reasonable energy densities can make flow batteries economically competent. Quinones are gaining interest as electrolytes for flow batteries in the past few years due to their ability to supply two electrons and impressive solubility characteristics, which result in relatively higher energy densities compared to vanadium redox couples. Suleyman et. al. [1] computationally explored the prospects of using Quinone molecules for both positive and negative electrolytes of RFB by substituting hydrogen atoms of Quinone molecules with different functional groups.

The present work focuses on identifying generic correlations to predict the solvation free-energy of Quinone derivatives [1] using the structural features of the molecules. One approach is to identify the solvation free-energy through group contribution method. Group contribution methods involve two steps. The contribution of each fragment of a structure is first identified using linear regression from the available data. Then the fragments of the given test molecule is counted and the end property is evaluated. Another approach is to predict solvation free-energy using Quantitative Structural Property Relationship (QSPR) methods. QSPR approach involves three steps. First step is to identify the structural features that can affect solvation free-energy such as polar surface area, refractivity etc. Then a linear/nonlinear relationship between the structural features and end property is identified. Finally, the structural feature values for the given test molecule is obtained and the end property is evaluated using the identified relationship. These relationships help in further exploring novel Quinone derivatives having higher aqueous solubilities.

References:

[1] S. Er, C. Suh, M. P. Marshak, and A. Aspuru-Guzik, â€œComputational design of molecules for an all-quinone redox flow battery,â€ *Chem. Sci.*, vol. 6, no. 2, pp. 885â€“893, 2015.

532748 A Novel Sulfonated Aromatic Polymer Membrane with Different Pendant Groups for Vanadium Redox Flow Batteries (VRFBs)

Abstract:

A novel sulfonated aromatic polymer membrane with different pendant groups for vanadium redox flow batteries (VRFBs)

Tongshuai Wang[1], Junyoung Han[2], Kihyun Kim[2], Janice Lee[1], Chulsung Bae[2]*, and Sangil Kim[1]*

[1]Department of Chemical Engineering, University of Illinois at Chicago, Chicago, IL 60607, USA

[2]Department of Chemistry & Chemical Biology, Rensselaer Polytechnic Institute, Troy, NY 12180, USA

Vanadium redox flow batteries (VRFBs) have received extensive attention due to its attractive features for the large-scale energy storage because of their design flexibility, high energy efficiency, absence of intercalation/deintercalation and stress build-up in electrodes, and safe operation. One of the most critical components in VRFB is an ion exchange membrane (IEM) which can prevent cross mixing of the multivalence vanadium ions, while it still allows the transport of protons (or anions) to complete the circuit. However, the trade-off relationship of proton exchange membrane (PEM) limits the energy efficiency (EE) of the VRFB due to low coulombic efficiency (CE) caused by vanadium ion cross over through the membrane or high membrane resistances. Anion exchange membrane (AEM) suffers from poor chemical stability and low voltage efficiency due to low ion conductivity.

In this study, we synthesized a series of novel sulfonated aromatic polymer membranes having high proton conductivity as well as proton selectivity. Vanadium/hydronium ion permeability, proton conductivity, area resistance, and ion exchange capacities of the membranes were measured. The single cell VRFB performance with the prepared membranes was evaluated and compared with the one using a commercial Nafion 212 membrane. The effect of the pendent group structure on the membrane ion selectivity and battery performance was also discussed.

It was found that the incorporation of aromatic structures on the pendent groups effectively improved the ion selectivity of membranes which can overcome the limitations of the trade-off relationship between the proton conductivity and selectivity. The membrane with aromatic pendent groups (BP-Ar-SH) shows 7 times higher vanadium/proton ion selectivity than that without aromatic groups (BP-SH) (5.18 *vs.* 0.75) and 3 times higher than Nafion 212 (5.18 *vs.* 1.78). As a result, the VRB single cell equipped with the BP-AR-SH membrane shows significantly better performance in comparison to the cell with Nafion membranes, as high as 90.79% EE (*vs.* 79.96% EE of Nafion 212) at 40 mA/cm^2. The cycle stability test shows that efficiencies of the membrane remain almost unchanged after running more than 50 cycles. Our results strongly suggest that the BP-Ar-SH membrane may provide a breakthrough toward the development of high-efficiency VRFB systems.

537112 Understanding the Influence of Thermal Pretreatment on Carbon Paper Electrodes for Vanadium Redox Flow Batteries

Abstract:

Grid-scale energy storage is anticipated to play a critical role in the decarbonization of the electric power sector by enabling the reliable integration of intermittent renewable sources and increasing the efficiency of non-renewable energy processes [1]. Redox flow batteries (RFBs) are a promising electrochemical device for such applications due to their independent energy and power scaling and long operational lifetime [2]. Aqueous vanadium RFBs (VRBs) have been the most successful chemistry, but further cost reductions are needed for ubiquitous adoption. Vanadium redox reactions have sluggish kinetics on the carbon electrodes typically employed in VRBs which motivates the implementation of pretreatment strategies to increase electrochemically active surface area and introduce catalytically active functional groups onto the electrode surface [3]. While many pretreatments have been successfully demonstrated most studies are empirically-driven. Few efforts have focused on systematically correlating changes in electrode structure and surface chemistry to shifts in electrochemical performance.

To this end, we utilize a suite of *ex-situ* characterization techniques, coupled with flow cell testing, to explore structure-performance relations of thermally pretreated carbon paper electrodes. Specifically, we vary thermal pretreatment temperature from 400 °C, a commonly used temperature for VRB electrode treatments [3,4], to 500 °C, to quantify changes in the physicochemical properties of the electrode, and link these changes to shifts in VRB performance. We leverage microscopic, spectroscopic, and analytical techniques to determine how microstructural, morphological, and chemical properties of electrodes are impacted by thermal pretreatment. We then evaluate the effect of these properties on flow cell performance using polarization, electrochemical impedance spectroscopy, and cell cycling. We find that many factors contribute to the observed electrode performance, including hydrophilicity, electrochemical surface area, and surface chemistry, and that not all of these factors improve with increasing temperature. Consequently, while the best overall performance is achieved with pretreatment at 475 °C, a 26% increase in maximum power density over the base case of pretreatment at 400 °C, this enhancement is based on a balance of critical properties rather than a maximization of all.

Acknowledgements

The authors acknowledge the financial support of the Joint Center for Energy Storage Research, an Energy Innovation Hub funded by the United States Department of Energy. K.V.G acknowledges additional funding from the National Science Foundation Graduate Research Fellowship. A.F.C. acknowledges financial support from the Swiss National Science Foundation (P2EZP2_172183).

References

1. Z. Yang et al., *Chem. Rev.*, **111**, 3577–3613 (2011).
2. M. Skyllas-Kazacos et al., *J. Electrochem. Soc.*, **158**, R55 (2011).
3. B. Sun and M. Skyllas-Kazacos, *Electrochim. Acta*, **37**, 1253–1260 (1992).
4. M. H. Chakrabarti et al., *J. Power Sources*, **253**, 150–166 (2014).

540773 2018 Outlook for Energy: A View to 2040

Abstract:

The Outlook for Energy is ExxonMobil's global view of energy demand and supply through 2040. It also highlights the dual challenge of providing affordable energy to support prosperity while reducing environmental impacts – including the risks of climate change. We use the Outlook to help inform our long-term business strategies and investment plans.
The Outlook anticipates significant changes through 2040 across the world to boost living standards, reshape the use of energy, broaden access to abundant energy supplies, and accelerate decarbonization of the world's energy system to address the risks of climate change.

- Global energy needs will rise about 25 percent over the period to 2040
- Electricity demand will nearly double in non-OECD countries
- Worldwide electricity from solar and wind will increase about 400 percent
- Natural gas will expand its role, led by growth for electricity generation
- Growth in oil demand will be driven by commercial transportation and chemical needs
- Efficiency gains and growing use of less-carbon intensive energy sources will contribute to a nearly 45 percent decline in the carbon intensity of global GDP

Energy is fundamental to modern life, and as the world's population approaches 9 billion people in 2040, we are challenged to improve living standards everywhere. Understanding the factors that drive the world's energy needs – and likely choices to meet those needs – is the mission of the Outlook. By sharing the Outlook with the public, we hope to broaden that understanding. Energy matters to everyone, and we all play a role in shaping its future.

541543 Energy Decarbonisation Scenarios

Abstract:

With the energy sector being associated with 2/3 of the annual greenhouse gas emissions, decarbonizing the energy is a key step to reduce their level. The National Determined Contributions (NDC's) announced so far by the participating countries do not meet the objectives of the Paris-climate agreement, as far as reaching a peak in emissions in the near future or decreasing to a level that is compatible with the stated long-term targets.

Using different scenarios for energy demand and supply, options for energy sector decarbonisation are evaluated. Meeting the 2-degree target requires a major transformation in the power sector, efficiency improvements – specially in the buildings and in the industry, and a revolution in transport towards fuel and modal shifts. With a significant increase in renewable energy, system integration and optimization will also be key in reducing the cost of the energy transition. Meeting more ambitious climate goals ("well below 2 degrees") will push further the technological innovation and implementation requirements throughout the entire energy value chain. A scenario aiming at zero net-emissions by 2060 is presented, along with a comparison to the 2-degree case.

542625 Fundamental Research Needs to Advance Energy Technologies

Abstract:

The U. S. Department of Energy (DOE), Office of Science, Office of Basic Energy Sciences (BES) supports fundamental research in chemical and materials sciences to provide the foundations for new energy technologies and to support DOE missions in energy, environment, and national security. This presentation will discuss opportunities for fundamental research to impact DOE's energy mission "to catalyze the timely, material, and efficient transformation of the nation's energy system and secure U.S. leadership in energy technologies" with a focus on the way we generate, store and use energy nationally. I will provide an overview of BES strategic planning over the past decade that identified priority research directions for advancing energy applications, highlight key scientific advances in these areas, and discuss some future opportunities for modern science, particularly chemical sciences, to accelerate the transformation of the U. S. energy portfolio.

540189 Modelling of Multiphasic Behavior of Biodiesel Transesterification Operating Below Critical Conditions Using CO_2 as a Co-solvent with PC-SAFT EoS

Abstract:

Biodiesel is composed of fatty acid esters generated via methanolysis or ethanolysis of triglycerides. The preferred method for biodiesel production is transesterification of triglycerides with methanol or ethanol. Phase behavior plays a critical role in biodiesel transesterification. The reaction is generally carried out at high temperatures, pressures and high short alcohols to oil ratios, in order to increase the solubility of non-polar and polar phases.

The process can be performed uncatalyzed using supercritical conditions: temperatures of 250 - 400 °C and pressures of 19- 24 MPa with high methanol/ethanol to oil ratios of 40:1 molar. The severity of these conditions guarantees good reaction kinetics and the formation of a singular phase in the reactor. This process has recently received much attention in literature however it still presents challenges regarding the intensive energy requirements for the process and downstream separation of products from excess reactants.[1–6]

The catalyzed process has different approaches including acid and basic homogeneous catalysts and is the current commercial path to produce biodiesel.[7] however ongoing challenges regarding the presence of water, free fatty acid impurities, and extra cost associated with the separation of the biodiesel from glycerol are still present.[7,8]

Heterogeneous catalysts have also been investigated showing economical profitability, and provide a solution for the final catalyst separation.[9] They vary in nature from alkaline earth, titanium silicates, anion exchange resins and polymers,[10–14]

Carbon dioxide has been suggested as co-solvent with beneficial properties for biodiesel transesterification at supercritical conditions because: it is abundant in nature, it has been widely used in vegetable oil extraction, it has a rather mild critical value (31.1 °C and 7.39 MPa), and increases the solubility between alcohols and triglycerides favoring conversion[15,16].

Findings from Soh et al[17] showed triglycerides transesterification does not need to be done at supercritical conditions and can be coupled with heterogeneously catalyzed reactions. Carbon dioxide works as a co-solvent for both the polar and oleic phase. Expanding the two phases with CO_2 increases the solubilized triglycerides on the alcohol phase, which consequently encourages the transesterification reaction. Almost complete conversion was obtained in moderate reaction times (< 2 Hr). Also, temperatures around 95 °C and pressures of 9.5 MPa were enough to achieve 98% methyl oleate yield in a tri-phasic catalyzed reaction process. Benefitial transport properties

Despite the system's simplicity, some species present on the biodiesel reaction, namely methanol/ethanol and glycerol have a high degree of molecular interaction as hydrogen bonding, and relevant polar forces. Understanding the systems phase behavior development from a modelling perspective requires models that can capture the complexity of this interactions.

Perturbed Chain Statistically Association Fluid Theory (PC-SAFT) by Gross and Sadowski,[18,19] has become a widely used engineering resource due to its accurate prediction capabilities and has been successfully applied to many associative mixtures. Additionally PC-SAFT has been extended to polar compounds.[20–24] In particular Nguyen et al.[25,26] and Corazza et al.[27] looked into biodiesel related phase equilibria, parametrizing pure methyl- and ethyl-esters at different conditions and obtaining promising results for methanol, ethanol, glycerol, and biodiesel binary and ternary systems.

Up to seven molecular parameters describe each pure component phase and are fitted to saturation data. The equation of states is written as an expansion of the Helmholtz free energy segregated into a reference fluid and the main interaction forces- i.e. dispersion, hydrogen bonding, and polar forces. Binary interaction coefficients are then calculated using binary data to mitigate accumulated errors that result from averaging approximations on mixtures.

Modelling of the biodiesel reaction in the presence of carbon dioxide has focused on supercritical conditions and most of the work done has been performed with cubic equations of state. Modelling this reaction at milder conditions of pressure and temperature where multiple phases are present using SAFT like equations has not been done before.

Modelling results achieved accurate phase behavior representation for pure, binary and ternary systems including carbon dioxide using a polar version of PC-SAFT. Group contribution methods allowed generalization for methyl-esters and ethyl-esters modelling simplifying the required components basis and minimizing the number of parameters. Small errors were obtained using very low values of binary interaction coefficients (below 7%) for binary systems. Successful modelling for liquid-liquid and vapor-liquid-liquid regimes suggest the possibility to determine optimal conditions for transesterification performed in multiphasic reactors in presence of CO_2

The global aim of this study is to fully characterize biodiesel-related compounds with carbon dioxide as a co-solvent with the PC-SAFT, predict supercritical, and sub-critical phase behavior generating binary and ternary diagrams to describe interactions amongst species. These calculations could potentially suggest a new process design by estimating favorable thermodynamic conditions as pressure, temperature and reactants ratios for biodiesel production using carbon dioxide as a co-solvent in subcritical conditions based on their phase behavior equilibria.

Keywords: Biodiesel, PC-SAFT, Phase-Behavior, Carbon-Dioxide, Co-solvent

References:

1. Madras, G., Kolluru, C. & Kumar, R. Synthesis of biodiesel in supercritical fluids. *Fuel* **83,** 2029–2033 (2004).
2. Balat, M. Biodiesel Fuel Production from Vegetable Oils via Supercritical Ethanol Transesterification. *Energy Sources, Part A Recover. Util. Environ. Eff.* **30,** 429–440 (2008).
3. He, H., Wang, T. & Zhu, S. Continuous production of biodiesel fuel from vegetable oil using supercritical methanol process. *Fuel* **86,** 442–447 (2007).
4. He, H., Sun, S., Wang, T. & Zhu, S. Transesterification kinetics of soybean oil for production of biodiesel in supercritical methanol. *JAOCS, J. Am. Oil Chem. Soc.* **84,** 399–404 (2007).

5. Demirbas, A. Biodiesel from vegetable oils via transesterification in supercritical methanol. *Energy Convers. Manag.* **43,** 2349–2356 (2002).
6. Sawangkeaw, R., Bunyakiat, K. & Ngamprasertsith, S. A review of laboratory-scale research on lipid conversion to biodiesel with supercritical methanol (2001-2009). *J. Supercrit. Fluids* **55,** 1–13 (2010).
7. Pinnarat, T. & Savage, P. E. Assessment of noncatalytic biodiesel synthesis using supercritical reaction conditions. *Ind. Eng. Chem. Res.* **47,** 6801–6808 (2008).
8. Vicente, G., Martínez, M. & Aracil, J. Integrated biodiesel production: A comparison of different homogeneous catalysts systems. *Bioresour. Technol.* **92,** 297–305 (2004).
9. West, A. H., Posarac, D. & Ellis, N. Assessment of four biodiesel production processes using HYSYS.Plant. *Bioresour. Technol.* **99,** 6587–6601 (2008).
10. Schuchardt, U., Vargas, R. M. & Gelbard, G. Transesterification of soybean oil catalyzed by alkylguanidines heterogenized on different substituted polystyrenes. *J. Mol. Catal. A Chem.* **109,** 37–44 (1996).
11. Georgogianni, K. G., Katsoulidis, A. K., Pomonis, P. J., Manos, G. & Kontominas, M. G. Transesterification of rapeseed oil for the production of biodiesel using homogeneous and heterogeneous catalysis. *Fuel Process. Technol.* **90,** 1016–1022 (2009).
12. Gryglewicz, S. Rapeseed oil methyl esters preparation using heterogeneous catalysts. **70,** (1999).
13. Demirbas, a. Biodiesel from Vegetable Oils with MgO Catalytic Transesterification in Supercritical Methanol. *Energy Sources, Part A Recover. Util. Environ. Eff.* **30,** 1645–1651 (2008).
14. Shibasaki-Kitakawa, N. *et al.* Biodiesel production using anionic ion-exchange resin as heterogeneous catalyst. *Bioresour. Technol.* **98,** 416–421 (2007).
15. Maçaira, J., Santana, A., Recasens, F. & Angeles Larrayoz, M. Biodiesel production using supercritical methanol/carbon dioxide mixtures in a continuous reactor. *Fuel* **90,** 2280–2288 (2011).
16. García-Jarana, M. B. *et al.* Use of supercritical methanol/carbon dioxide mixtures for biodiesel production. *Korean J. Chem. Eng.* **33,** 2342–2349 (2016).
17. Soh, L., Curry, J., Beckman, E. J. & Zimmerman, J. B. Effect of System Conditions for Biodiesel Production via Transesterification Using Carbon Dioxide-Methanol Mixtures in the Presence of a Heterogeneous Catalyst. *ACS Sustain. Chem. Eng.* 387–395 (2014). doi:10.1021/sc400349g
18. Gross, J. & Sadowski, G. Perturbed-Chain SAFT: An Equation of State Based on a Pertubation Theory of Chain Molecules. *Ind. Eng. Chem. Res* 1244–1260 (2001).
19. Tumakaka, F., Gross, J. & Sadowski, G. Modeling of polymer phase equilibria using Perturbed-Chain SAFT. *Fluid Phase Equilib.* **194–197,** 541–551 (2002).
20. Gubbins, K. E. & Twu, C. H. Thermodynamics of polyatomic fluid mixtures—I theory. *Chem. Eng. Sci.* **33,** 863–878 (1978).
21. Twu, C. H. & Gubbins, K. E. Thermodynamics of Polyatomic Fluids Mixtures-II. *Chem. Eng. Sci.* **33,** 879–887 (1977).
22. Twu, C. H., Gubbins, K. E. & Gray, C. G. Thermodynamics of mixtures of nonspherical molecules. III. Fluid phase equilibria and critical loci. *J. Chem. Phys.* **64,** 5186–5197 (1976).
23. Vrabec, J. & Gross, J. Vapor - Liquid Equilibria Simulation and an Equation of State Contribution for Dipole - Quadrupole Interactions. 51–60 (2008).
24. K. Jog, W. G. Chapman, P. Application of Wertheim's thermodynamic perturbation theory to dipolar hard sphere chains. *Mol. Phys.* **97,** 307–319 (1999).
25. Nguyen-Huynh, D., Passarello, J.-P., Tobaly, P. & de Hemptinne, J.-C. Modeling Phase Equilibria of Asymmetric Mixtures Using a Group-Contribution SAFT (GC-SAFT) with a k ij Correlation Method Based on London's Theory. 1. Application to CO 2 + n -Alkane, Methane + n -Alkane, and Ethane + n -Alkane Systems. *Ind. Eng. Chem. Res.* **47,** 8847–8858 (2008).
26. Nguyen Huynh, D., Nguyen Thi, T. & Van Dinh, S. T. Predicting the temperature / pressure dependent density of biodieselfuels. *Petrovietnam* **10,** 46–58 (2012).
27. Corazza, M. L., Fouad, W. A. & Chapman, W. G. PC-SAFT predictions of VLE and LLE of systems related to biodiesel production. *Fluid Phase Equilib.* **416,** 130–137 (2016).

527201 High Pressure Hydrogen Generation and Simultaneous Carbon Dioxide Separation with Chemical Looping Hydrogen

Abstract:

High pressure hydrogen generation with inherent carbon dioxide separation via fixed bed chemical looping

Decentralised production of hydrogen from renewable primary energy sources provides the basis for a sustainable energy system with low emissions. At Graz University of Technology the reformer steam-iron cycle (RESC) was developed for the decentralized production of pure pressurized hydrogen [1]. Synthesis gas from renewable feedstock, e.g. biogas or gasified biomass, is used in a chemical looping fixed-bed system to release pure, pressurized hydrogen with efficiencies of up of 75 % [2],[3]. This process additionally enables the inherent separation of carbon dioxide from carbonaceous feedstock to facilitate carbon capture and storage applications. Literature data showed that the pre-pressurized release of hydrogen with 100 bar could significantly reduce the energy demand for compression of up to 80% [4].

The present work proves the possibility of producing pure, pressurized hydrogen while inherently separating a carbon dioxide stream with fixed-bed chemical looping [5]. The experiments were conducted in a tubular fixed bed reactor, which is part of a modified test system for catalyst characterization (Figure 1). The reactor was filled with oxygen carrier material in powder form with the composition of 95% Fe_2O_3 + 5% Al_2O_3, synthesized by wet chemical impregnation method [6]. Simulated syngas mixtures from methane steam reforming at steam to carbon ratios from 1.4 to 2.0 were used to reduce the oxygen carrier. The gas quality during all experiments was measured continuously with a Micro Gas Chromatograph (Inficon Fusion). Partial reduction led to a pure carbon dioxide offgas stream. The subsequent steam oxidation enabled a direct pre-pressurized hydrogen withdrawal at 30 bars with a hydrogen purity of 99.3%.

To investigate the maximum possible hydrogen outlet pressure of the system an additional experimental series was conducted to investigate the possibility of high-pressure steam oxidation. Within the experiments, pressurized steam oxidations with release pressures of up to 95 bar and hydrogen purities up to 99.9% could be achieved. XRD, SEM, light microscopy and BET measurements were used to characterize the oxygen carrier before and after the experiments. The results showed no degradation of the oxygen exchange capacity over 20 high pressure reduction and oxidation cycles.

Acknowledgement

The authors gratefully acknowledge financial support by the Austrian Federal Ministry of Transport, Innovation and Technology (BMVIT), Austrian Federal Ministry of Education, Science and Research (BMBFW) and the Austrian Research Promotion Agency (FFG) through the energy research program.

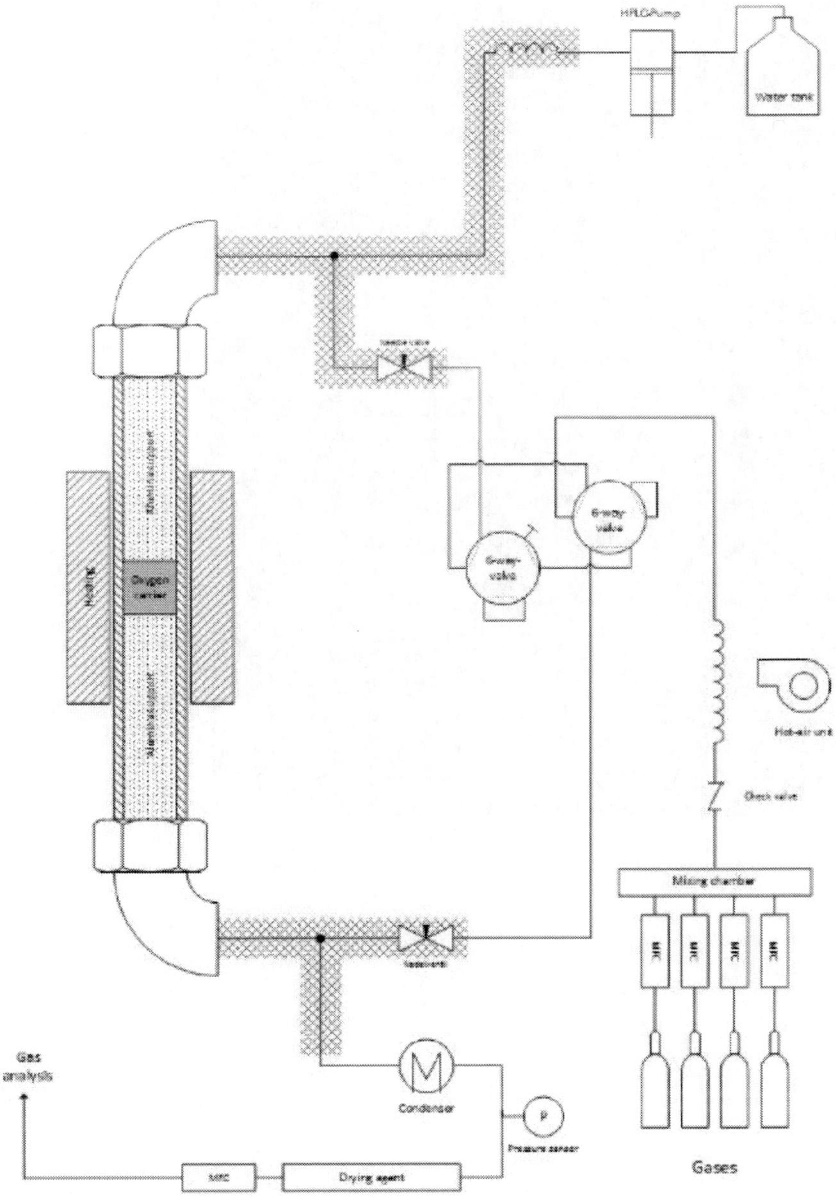

Figure 1: Scheme of the reactor system for pressures up to 100 bar

SEM HV: 20.0 kV
WD: 15.00 mm
VEGA3 TESCAN
SEM MAG: 5.00 kx
Det: SE
10 µm

Figure 2: SEM and Light microscopy pictures of the oxygen carrier

[1] V. Hacker, "A novel process for stationary hydrogen production : the reformer sponge iron cycle (RESC)," *J. Power Sources*, vol. 118, pp. 311–314, 2003.

[2] G. Voitic, S. Nestl, K. Malli, J. Wagner, B. Bitschnau, F.-A. Mautner, and V. Hacker, "High purity pressurised hydrogen production from syngas by the steam-iron process," *RSC Adv.*, vol. 6, no. 58, 2016.

[3] S. D. Fraser, M. Monsberger, and V. Hacker, "A thermodynamic analysis of the reformer sponge iron cycle," *J. Power Sources*, vol. 161, no. 1, pp. 420–431, 2006.

[4] M. Ozsaban and A. Midilli, "A parametric study on exergetic sustainability aspects of high-pressure hydrogen gas compression," *Int. J. Hydrogen Energy*, vol. 41, no. 11, pp. 5321–5334, 2016.

[5] R. Zacharias, S. Bock, and V. Hacker, "High pressure hydrogen generation with inherent carbon dioxide separation via fixed bed chemical looping.," *in preperation*.

[6] C. D. Bohn, J. P. Cleeton, C. R. Müller, S. Y. Chuang, S. A. Scott, J. S. Dennis, C. R. Muller, S. Y. Chuang, S. A. Scott, and J. S. Dennis, "Stabilizing Iron Oxide Used in Cycles of Reduction and Oxidation for Hydrogen Production," *Energy & Fuels*, vol. 24, no. 7, pp. 4025–4033, Jul. 2010.

530059 Renewable Transport Fuels, Heat and Electricity from Miscanthus: Optimisation for Design, Planning and Operation of Sustainable Value Chains

Abstract:

Biofuels and bioenergy (i.e. bio-heat and bio-electricity) can play a prominent role in decarbonising the energy system. Biofuels are compatible with existing vehicles and transport fuel supply infrastructure and they are one of the most affordable alternatives to fossil fuel. Currently, they are the only viable replacement fuels for the aviation and shipping sectors since hydrogen fuel cells and electric batteries have not yet advanced sufficiently to power aircrafts and ships. Further, production of heat and electricity from biomass is a mature and proven low-carbon technology, which when combined with carbon capture and storage can result in negative GHG emissions. However, biofuels and bioenergy are one of the most complex and controversial alternative sources for energy. If not properly planned and managed, they can heighten competition with food production and cause detrimental effects on the environment and biodiversity. Therefore, their supply chains need to be carefully designed, planned and operated in order to ensure that their production and utilisation are sustainable, do not contribute to global warming and do not result in water and land use conflicts. A biomass supply chain typically consists of growing and harvesting biomass feedstocks, pre-processing and pre-treatment, conversion to intermediates, final products and energy services, logistics and inventory and waste management.

There has been a lot of research into biomass supply chain modelling and optimisation. However, most them considered production of bio-ethanol from arable crops, which can cause competition with food production, as well as being water and energy-intensive. In Great Britain (GB), energy crop such as Miscanthus × giganteus (hereafter Miscanthus) are promising alternatives to arable crops as feedstocks for biofuel and bioenergy production. Miscanthus is a perennial grass that can be cultivated in marginal and degraded land, which is not used for food production; it requires low input (in terms of energy, water and fertilisers); it is fast-growing and high yielding and once established it can provide biomass feedstock for up to 20 years without the need for replanting. To date, no optimisation model has been developed for biofuel and bioenergy supply chains based on Miscanthus.

The aim of this project is to examine the large scale utilisation of Miscanthus to satisfy transport fuel, electricity and heat demands in GB. A mixed integer linear programming (MILP) model is developed that can simultaneously determine the design, planning and operation of the supply chains. GB is divided into 50 km × 50 km squares, in order to capture the suitable land for Miscanthus cultivation and the variation in productivity/yield, distribution of product demands, as well as to model the transport of resources. To avoid competition with food production, only grassland in GB is considered. GIS modelling is performed to exclude grassland areas that are not suitable for cultivation and harvesting based on elevation, slope and soil organic carbon, as well as those that do not comply with a number of environmental and social constraints (e.g. protected areas, urban areas etc.). Different conversion technologies, at different scales, are modelled including pelletising, torrefaction, gasification, Fischer-Tropsch (FT) synthesis, di-methyl ether (DME) synthesis, lignocellulosic fermentation, pyrolysis and pyrolysis oil upgrading, and integrated gasification combined cycle (IGCC) for combined heat and power generation. The valuable products in the supply chain include: FT-jet (aviation fuel); upgraded pyrolysis oil (UPO, marine fuel); DME, FT-diesel and ethanol (road transport fuel); and heat and electricity. The spatial distributions of the demands for these products across GB are modelled. The planning horizon considered is from 2018 to 2050.The optimisation determines where to grow Miscanthus, what technologies to invest in and where to locate them and which products to produce in order to maximise the net present value (NPV) of the supply chains. It also decides between centralised and distributed processing or a combination of the two; whether to densify or pre-process the biomass before transportation; how the biomass is transported to the processing facilities and the products distributed to customers; and how to manage inventory to balance Miscanthus availability and demands for products. In this conference, we will present the optimisation model as well the sustainable and profitable Miscanthus supply chains identified using the model.

References:

[1] S. Samsatli, N.J. Samsatli (2018). A multi-objective MILP model for the design and operation of future integrated multi-vector energy networks capturing detailed spatio-temporal dependencies. Applied Energy. DOI: 10.1016/j.apenergy.2017.09.055.

[2] S. Samsatli, N.J. Samsatli, N. Shah (2015). BVCM: a comprehensive and flexible toolkit for whole-system biomass value chain analysis and optimisation - mathematical formulation. Applied Energy, 147, pp. 131-160.

[3] S.M. Jarvis, S. Samsatli (2018). Technologies and infrastructures underpinning future CO2 value chains: a comprehensive review and comparative analysis. Renewable & Sustainable Energy Reviews, 85, 46-68.

[4] S. Samsatli, N.J. Samsatli (2015). A general spatio-temporal model of energy systems with a detailed account of transport and storage. Computers and Chemical Engineering, 80, 155-176.

532392 Gasification of Glucose in the Reactor CREC-Riser Simulator: Stable Ni-(La or Ce)/g-Alumina Catalysts and Effect of Reaction Time and Steam/Glucose

Abstract:

Glucose Gasification in the CREC-Riser Simulator Reactor Using a Ni-(La or Ce)/g-Alumina Fluidizable Catalyst. Effect of Reaction Time and Steam/Glucose Ratio.

A. Sánchez Enríquez[1], D.G. González Castañeda[1], I. Cruz Reyes[1], A.R. Calzada Hernandez[1], H.I. de Lasa[2], B. Serrano Rosales.[*1,]

[1] *Universidad Autónoma de Zacatecas, Unidades Académicas de Ingeniería Eléctrica y Ciencias Químicas, Campus UAZ Siglo XXI, Carr, a Guadalajara km 6, Ejido La Escondida, Zacatecas, Zac.,98160.*

[2] *Chemical Reactor Engineering Centre, Department of Chemical and Biochemical Engineering, The University of Western Ontario, London, ON N6A 5B9, Canada.*

(*) beniser@prodigy.net.mx

Keywords: Hydrogen, Glucose, Nickel, Cerium, Lanthanum, Gasification

<>1. Introduction

The gradual depletion of fossil fuels requires the development of alternative technologies such as biomass gasification [1]. Gasification involves a complex network of heterogeneous reactions, with the use of catalysts which significantly enhance the gasification reaction network.

Nickel catalysts have been reported as active for methane and heavy hydrocarbon conversion[2]. Biomass gasification may require a new class of fluidizable catalytic materials to achieve the desired syngas composition with minimum amounts of tars. To accomplish this, a 5wt% nickel catalyst supported on fluidizable g-Al_2O_3 is proposed in the present study. Furthermore, the influence of lanthanum and cerium as added promoters, is also examined. Catalysts are synthesized using a direct reduction method without catalyst precursor calcination. Various prepared fluidizable catalysts are evaluated by changing the operating parameters such as reaction time and steam/glucose ratio. Furthermore, the influence of oxidation/reduction cycles on catalytic activity is also thoroughly examined. <>2. Experimental Methodology

Catalyst Preparation. - The 5%Ni/g-Al_2O_3 catalysts were prepared with a pH of 1 and 4. An incipient wetness impregnation was followed by direct precursor reduction as follows[3]: a) The alumina support sample was dried during 12 hours at 110°C, b) A $Ni(NO_3)$ ·6H O was added to a deionized water solution, c) The alumina support was placed under a vacuum, for 20 minutes, d) The $Ni(NO_3)$ solution was added slowly (drop-by-drop) to the alumina, e) The resulting paste was dried during 12 hours, e) The nickel impregnated alumina was reduced in a fixed bed using a 160 cm^3/min hydrogen flow and a 5°C/min heating ramp. The temperature was held at 260°C for 1 hour, f) Following this, a second 5°C/min heating ramp was used until 480°C was reached. This temperature was held for 10 hours, g) Finally, the catalyst sample was cooled down to room temperature. A similar procedure was used for the catalyst promoted with Ce and La, with A and B representing 0.5wt%, 1wt% and 2wt% for Ce and La, respectively. In the case Ce and La, the promoter addition step b) was modified, using Ni (NO) ·6H O and La or Ce nitrate. These two nitrates were mixed together in deionized water solution and added to the support in a co-impregnation process [3]. Based on this methodology, fourteen catalysts were prepared. These catalysts were characterized using XRD, AA, BET, PSD and SEM.

Gasification in the CREC – Riser Simulator Reactor. – The CREC Riser Simulator is a laboratory scale fluidized bed [4], with a 53 cm^3 volume. The temperature in these experiments was 600°C. A glucose solution was injected into the reactor with and without a catalyst. The reaction times were 5, 10, 20, 30 and 40 seconds. The amount of catalyst used was 0.0500 g. Ratios of steam/glucose (S/B) of 0.5, 1, and 1.5 g/g were used under an inert atmosphere of argon (1 atm). The samples were analyzed in a GC-AT 7890A, with TCD and FID detectors. All the catalysts were regenerated first with air (15 min., P = 20 psia) and later with hydrogen (15 min., P = 20 psia). This was done before each injection. However, to test catalyst regeneration effectiveness, two additional steps were included: a) Prior to the runs catalyst was oxidized with air, with reduction being omitted, b) Before experiments catalyst was not either oxidized or reduced. Operational conditions were selected based on the results from the already completed runs obtained by the ANOVA analysis. <>3. Results and discussion

As a first step and as a reference, runs were carried out at 600 °C without the catalyst being loaded in the catalyst basket. These experiments were designated as thermal runs and the results were compared to the ones which used the 5%Ni on g-Al_2O_3 catalyst, with A=0.5, 1 and 2%.

Figure 1 reports the H_2, CO, CO_2 and CH_4 molar fractions obtained during the thermal runs and compare with the syngas molar fractions for three of best catalysts as follows: a) 5%Ni/g-Al_2O_3(pH4), b) 5%Ni-2%La/g-Al_2O_3(pH1), c) 5%Ni-2%Ce/g-Al_2O_3(pH4). Figure 2 reports the ratios of H_2/CH_4 for the S/B = 0.5, 1 and 1.5.

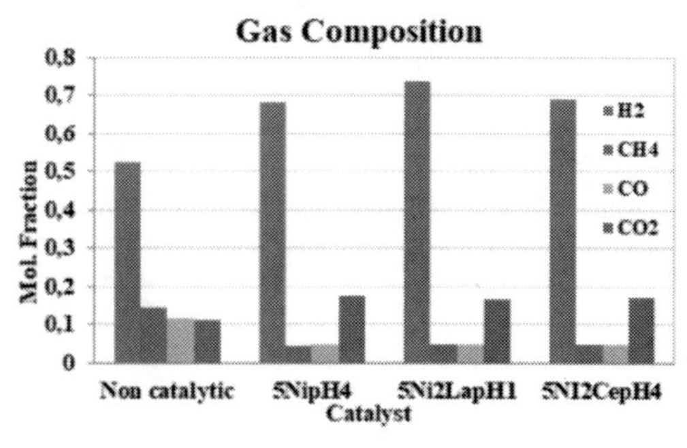

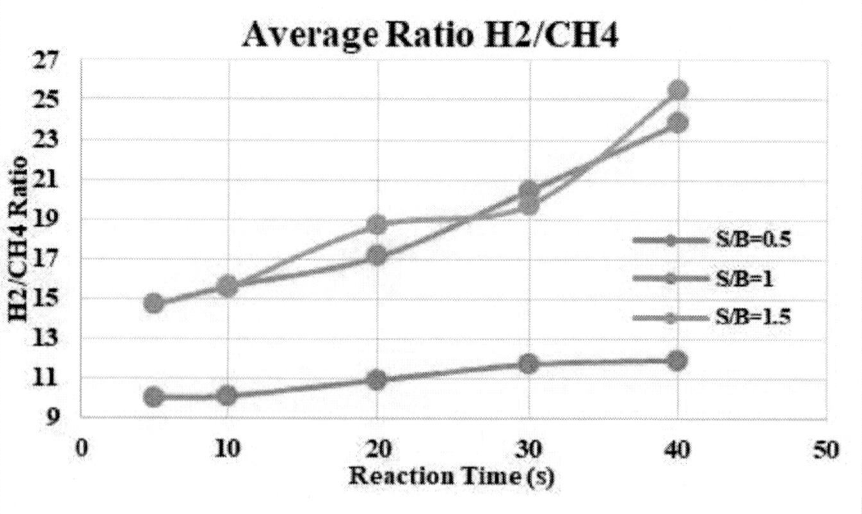

Figure 1.- Synthesis gas molar fractions for: a) the thermal case, b) a 5%Ni(pH4) catalyst, c) a 5%Ni-2%La(Ph1) catalyst, d) a 5%Ni-2%Ce(pH4) catalyst. Reaction time: 20 s, Steam/B= 1, Cat/B=3.2, Mass of Catalyst: 50 mg, T=600 °C.	Figure 2.- Effects of Reaction Time and Steam/B on H_2/CH_4 Ratio for 5%Ni-2%La/g-Al_2O_3(pH1). Cat/B=3.2, Mass of Catalyst: 50 mg, T=600 °C.

Figure 1 shows that when using the 5%Ni(pH4) catalyst, hydrogen production is increased by 32 % and this while compared to the thermal runs. As well, when using a 5%Ni-2%La(pH1) catalyst, there is an additional hydrogen increase of 10%. This demonstrates the positive effect of both Ni and La, on catalytic glucose conversion into H_2. In fact, the performance of the 5%Ni-2%La/g- Al_2O_3 (pH1) catalyst 0.73 H_2 molar fraction suggests that there is enhanced steam reforming, methane dry reforming and water gas shift reactions during glucose gasification.

Figure 2 reports the effects of both reaction time and S/B on the H_2/CH_4 ratio for the best performing 5%Ni-2%La/g-Al_2O_3(pH1) catalyst. It is observed that for all the S/B values, the H_2/CH_4 ratio augments with reaction time. This effect is particularly noticeable for the higher S/B ratios, with hydrogen molar fractions increasing and the methane (a primary gasification product) molar fraction being reduced.

Figure 3 shows the synthesis gas molar fractions for the 5%Ni-2%La/g-Al_2O_3 (pH1) catalyst and their changes with consecutive reaction cycle number. One can observe that for the freshly reduced catalyst, the first injections led to higher H_2. Hydrogen molar fractions decreased slightly however, for the second and subsequent injections, remaining close to constant. This consistent trend was observed (not shown) for the various catalysts of this study. It thus, appears that oxidization (regeneration) between injections was, in principle, fully adequate in removing any possible formed coke.

Furthermore, and to further study catalyst stability, 6 consecutive injections were used with the 5%Ni-2%La/g-Al_2O_3(pH1) catalyst, without regeneration in between injections, as reported in Figure 4. One can notice that the results in Figure 4 are similar to the ones in Figure 3. This shows that the 5%Ni-2%La/g-Al_2O_3(pH1) catalyst regeneration before each cycle may not be needed. Based on this, one can speculate that oxidation-reduction processes coexist in the 5%Ni-2%La/g-Al_2O_3(pH1) catalyst during every glucose/steam run [5].

As a result, it is anticipated that these efficient and stable catalysts could simplify the configuration of the industrial gasification plant, without the need of a catalyst regeneration unit.

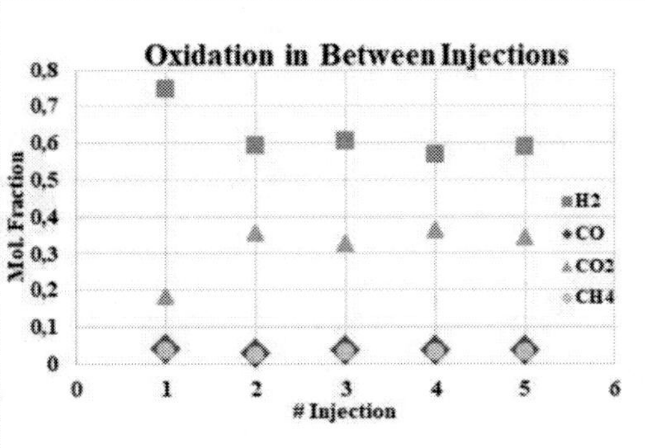

	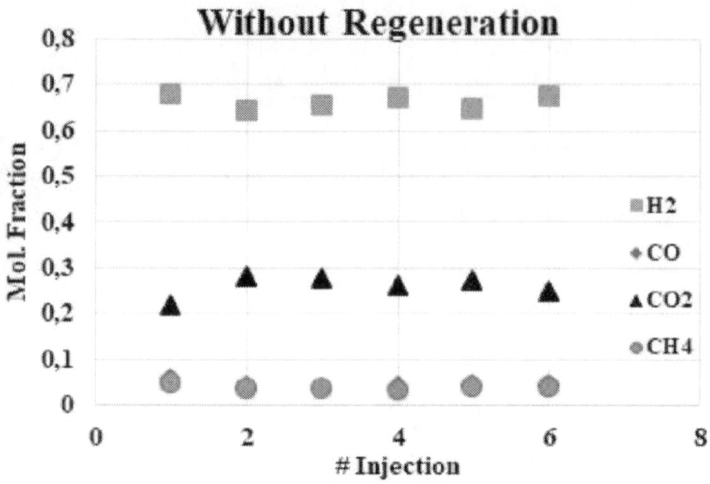
Figure 3.- Molar fraction for the 5%Ni-2%La/g-Al_2O_3(pH1) catalyst, at Steam/B =1, Time = 20 s, Cat/B=3.2, and the mass of catalyst being 50 mg, T=600 °C.	Figure 4. Molar fraction for the 5%Ni-2%La/g-Al_2O_3(pH1) catalyst, without regeneration, T = 600°C.

<>4. Conclusions

1. A 5%Ni-2%La/g-Al_2O_3(pH1) catalyst was successfully prepared for H_2 production. The performance of this catalyst was demonstrated in a CREC Riser Simulator.

2. It is shown that the prepared catalyst displayed significant improvements (40%) in the hydrogen molar fractions, and this when compared to those obtained during thermal runs.

3. It is proven that the synthesized catalyst showed a significantly reduced methane molar fraction when compared to that observed in the thermal experiments.

4. It is demonstrated that the performance of the prepared catalyst is favorably affected by increasing both S/B rations and reaction times.

5. It is shown that oxidation-reduction processes take place concurrently during glucose gasification with this leading to stable hydrogen production.

References

[1] J. Mazumder, H. I. de Lasa, *Chemical Engineering Journal* 293 (2016) 232-242.

[2] D. Sutton, B. Kelleher, J.R.H. Ross, *Fuel Processing Technology* 73.3 (2001) 155-173.

[3] C. H. Bartholomew, R. J. Farrauto, *Journal of Catalysis* 45.1 (1976) 41-53.

[4] H.I. de Lasa, Riser Simulator, US Patent 5102628, (1992).

[5] Juan-Juan, Román Martínez, Illán Gomez, *Applied Catalysis A: General*, 355 (2009) 27-32.

535194 Effect of Stirring on the Hydrate Formation Rate for Natural Gas Storage and Transportation Perspective

Abstract:

The liquefaction of natural gas (LNG) is an energy intensive and complex process, which requires the cooling of natural gas up to -161 °C. The technology required for conversion of natural gas to LNG is also sophisticated and requires large sum of investments. Therefore, the interest of academic and industrial researchers is shifting towards the use of hydrates as the mode of natural gas transportation and storage.

The process for transporting natural gas in the form of hydrates requires the conversion of natural gas into hydrate bulky crystals by cooling natural gas until the optimum temperature and then refrigerating at the suitable temperature of about -10 °C. This process compared to LNG seems to be simple and less energy intensive. The hydrate formation rate is the key when it comes to the conversion of natural gas to the hydrates. The mechanical factors like stirring rate or agitation of gas-water mixture can affect the conversion of natural gas into hydrates. Therefore, in this work the effect of stirring on the hydrate formation rate has been evaluated at the varying stirring rates (100-1400 RPM).

All the experimental work has been conducted using the high pressure cell that consist of a magnetic stirrer and using a gas mixture that consist of four components such as methane, ethane, carbon dioxide and nitrogen. The experimental work has been conducted at the high pressure of 98 bars and variable stirring rates (100-1400 RPM). The experimental results indicate that there exists a threshold limit above and below which no hydrate formation occurs. This shows that optimum stirring rate is necessary to facilitate the conversion of natural gas into the hydrates. This work provides a valuable insight to the researcher working in the area of natural gas storage and transportation. It is also valuable for the process engineering experts and professionals working in the area of energy transfer and storage.

Acknowledgment

This work was made possible by GSRA # 2-1-0603-14012 from the Qatar National Research Fund (a member of Qatar Foundation). The statements made herein are solely the responsibility of the authors.

MgO/Mg based Solar Driven CH_4 Reforming and H_2O Splitting Process

(Principal Author)
Rahul Bhosale
Department of Chemical Engineering
College of Engineering
Qatar University, Doha, Qatar
rahul.bhosale@qu.edu.qa

(Co-Authors)
Fares AlMomani
Department of Chemical Engineering
College of Engineering
Qatar University, Doha, Qatar

Prepared for Presentation at
American Institute of Chemical Engineers
2018 Fall Meeting
Pittsburgh, PA, USA
October 28 – November 2, 2018

MgO/Mg based Solar Driven CH_4 Reforming and H_2O Splitting Process

Rahul R. Bhosale*, Fares AlMomani

Department of Chemical Engineering
College of Engineering
Qatar University, Doha, Qatar

Keywords: Magnesium oxide, CH_4 reforming, Water splitting, H2, Syngas

Abstract

In this paper, MgO/Mg based solar driven methane reforming and water-splitting process was thermodynamically investigated for the co-production of H_2 and syngas. The equilibrium thermodynamic analysis was performed using the HSC Chemistry software and its thermodynamic database. These preliminary results provides the basis for a detailed solar reactor efficiency analysis of the MgO/Mg based solar thermochemical methane reduction and water splitting process.

1. Introduction

To fulfil the future energy demand, solar fuels such as solar H_2 or solar syngas (useful as a precursor for the production of liquid transportation fuels) are considered as potential options. Metal oxide based solar driven thermochemical H_2O/CO_2 splitting cycle is investigated significantly by plenty of research teams all over the world [1-46]. The prime limitation of the metal oxide based thermochemical cycle is the need of high temperature for thermal reduction. The requirement of high temperature yield into difficulties in the reactor design and reduces the process efficiency significantly.

The modification of the metal oxide based thermochemical cycle by employing methane as the reducing agent in the first step provides a promising option to reduce the thermal reduction temperature. The reaction chemistry associated with the solar driven methane reforming and water-splitting process is explained below:

$$MO + CH_4 \rightarrow MO_{red} + H_2 + CO \qquad (1)$$

$$MO_{red} + H_2O/CO_2 \rightarrow MO + H_2/CO \qquad (2)$$

In this paper, as an initial attempt, the MgO/Mg based solar driven methane reforming and water-splitting process was thermodynamically investigated and the preliminary results are explained in detail.

2. Equilibrium Thermodynamic Analysis

The thermodynamic analysis shows (based on the equibrium compositions associated with the thermal reduction of MgO in absence of any inert gas or methane as a reducing agent) that the direct thermal decomposition of MgO is not possible below 3680 K.

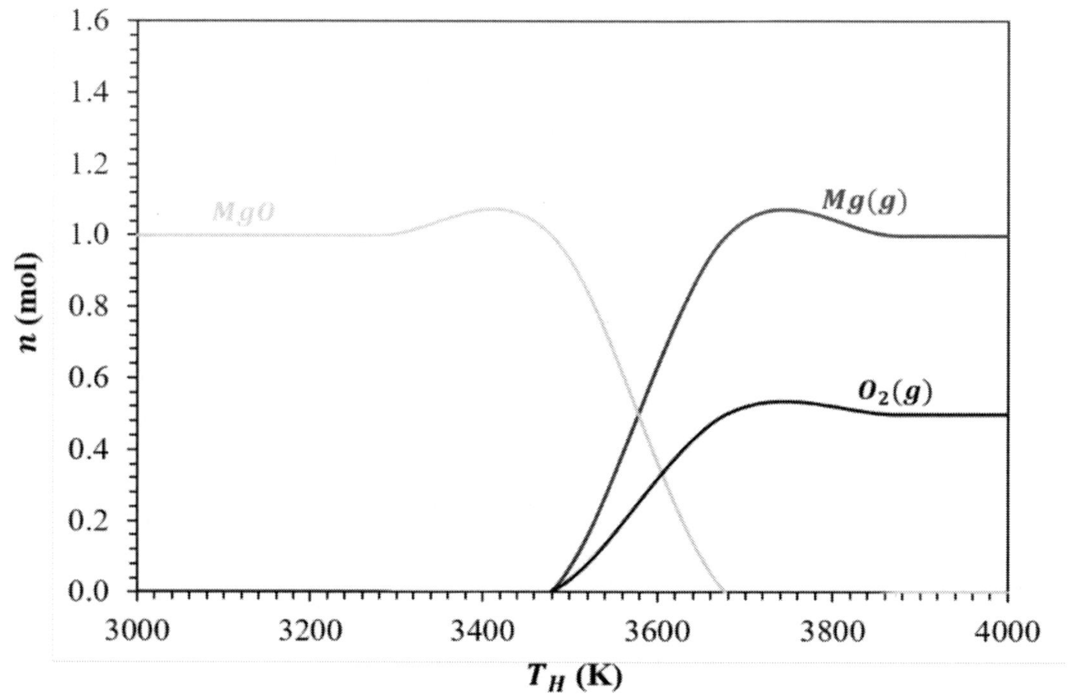

Fig. 1. Direct thermal decomposition of MgO in absence of Ar or CH_4.

In contrast, due to the inclusion of CH_4 as the reducing agent, the thermal reduction of MgO is possible below 2100 K. Furthermore, the production of syngas (combined stream of H_2 and CO with molar ratio equal to 2) is also possible due to the methanothermal reduction of MgO. Due to the increase in the molar ratio of CH_4/MgO from 0.1 to 1, the production of H_2 and CO increases significantly. In all cases, the complete thermal reduction of MgO is feasible at around 2080 K. The production of H2 starts in the range of 450 to 500 K and reaches its peak value 1640 K. Carbon formation also starts at around 450 to 500 K and suppresses to nullified quantity at around 2080 K. CO formation begins at around 1490 K and reaches its maxima by 2080 K. As the complete thermal reduction of MgO, formation of H2 and CO with maximum yield and no carbon formation is feasible above 2080 K, all the solar efficiency calculations will be performed at 2100 K (lowest possible temperature with most desirable results). Sample plots associated with the methanothermal reduction of MgO are listed below.

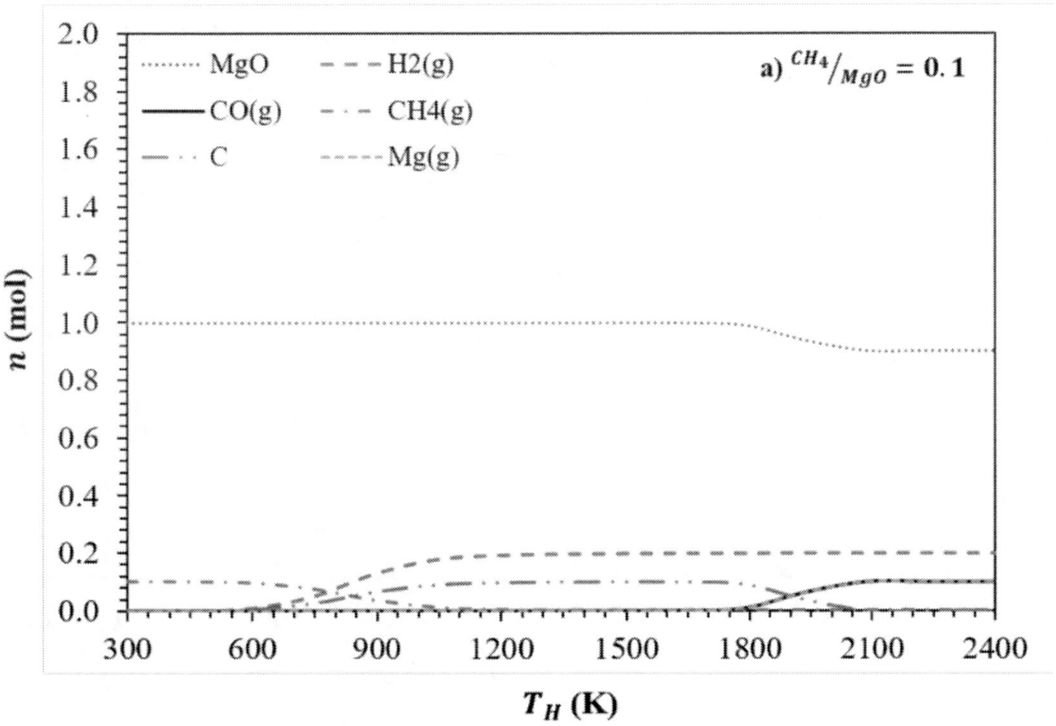
MgO
H2(g)
CO(g)
CH4(g)
C
Mg(g)
a) $^{CH_4}/_{MgO} = 0.1$
n (mol)
T_H (K)
2.0
1.8
1.6
1.4
1.2
1.0
0.8
0.6
0.4
0.2
0.0
300
600
900
1200
1500
1800
2100
2400

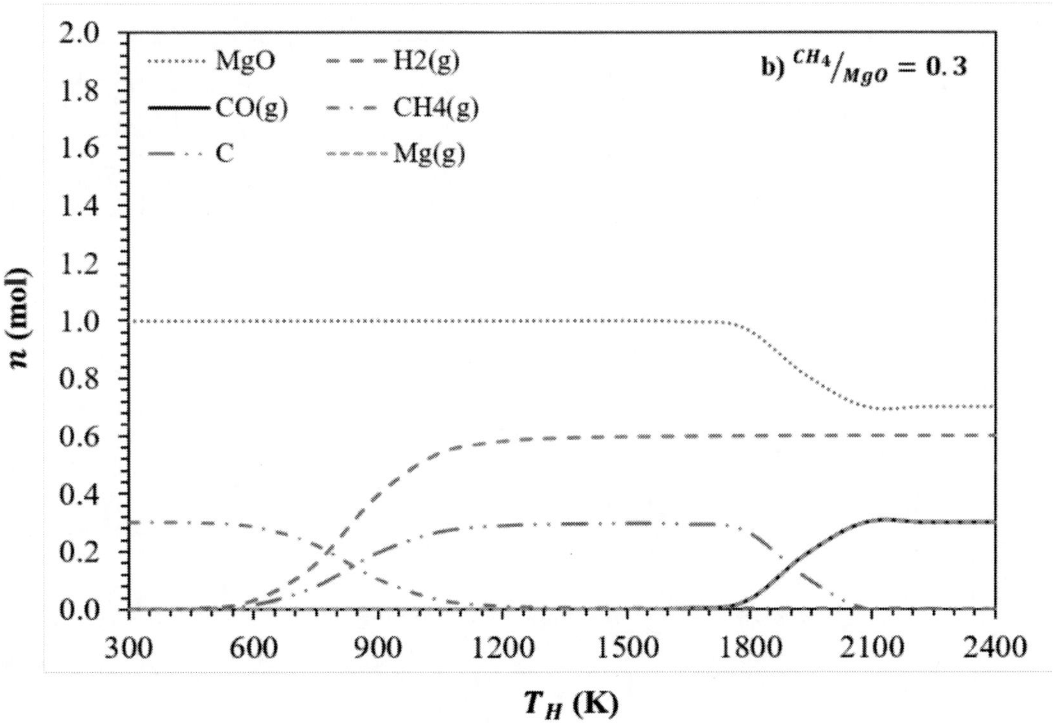
MgO
H2(g)
CO(g)
CH4(g)
C
Mg(g)
b) $^{CH_4}/_{MgO} = 0.3$
n (mol)
T_H (K)
2.0
1.8
1.6
1.4
1.2
1.0
0.8
0.6
0.4
0.2
0.0
300
600
900
1200
1500
1800
2100
2400

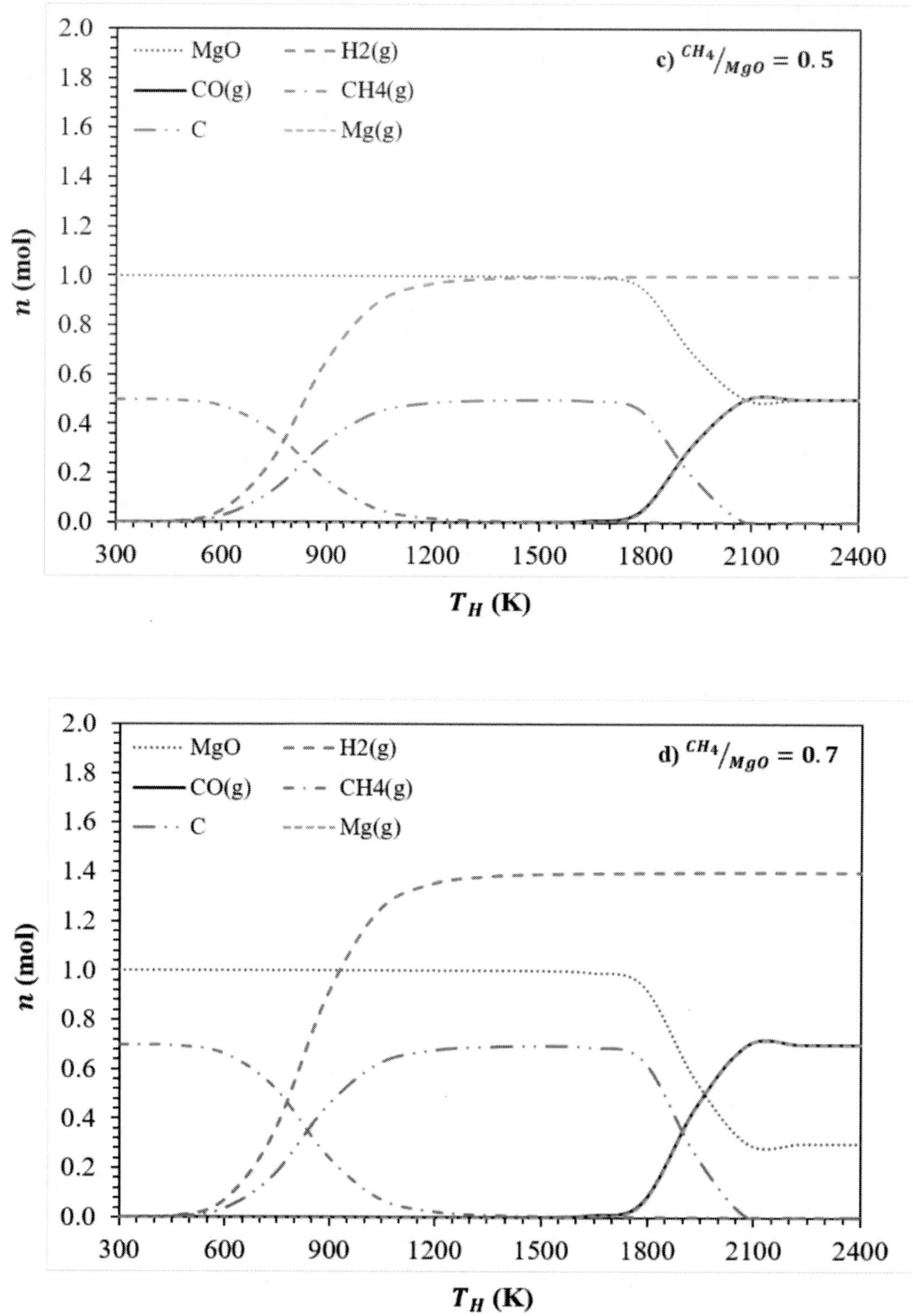

Fig. 2. Thermal decomposition of MgO in presence of CH_4.

3. Conclusions

The equilibrium thermodynamic analysis of the solar driven MgO/Mg based methane reforming and water splitting process indicates that the thermal reduction temperature required for the complete dissociation of MgO into Mg (with maximum production of H2 and CO) can be reduced by 1600 K due to the inclusion methane as a reducing agent.

4. Acknowledgments

This publication was made possible by the NPRP grant (NPRP8-370-2–154) from the Qatar National Research Fund (a member of Qatar Foundation). The statements made herein are solely the responsibility of author(s).

References

[1] J.R. Scheffe, A. Steinfeld, Oxygen exchange materials for solar thermochemical splitting of H_2O and CO_2: a review, Materials Today 17 (2014) 341–348.
[2] R. R. Bhosale, G. D. Takalkar, P. N. Sutar, A. Kumar, F. AlMomani, M. Khraisheh, A decade of ceria based solar thermochemical H_2O/CO_2 splitting cycle, International Journal of Hydrogen Energy (2018) Article In-Press.
[3] D. Yadav, R. Banerjee, A review of solar thermochemical processes, Renewable and Sustainable Energy Reviews 54 (2016) 497 – 532.[4] R. R. Bhosale, A. Kumar, F. AlMomani, R. B. Gupta, Solar thermochemical $ZnO/ZnSO_4$ water splitting cycle for hydrogen production, International Journal of Hydrogen Energy 43 (2017) 23474-23483.
[5] P.G. Loutzenhiser, A. Steinfeld, Solar syngas production from CO2 and H2O in a two-step thermochemical cycle via Zn/ZnO redox reactions: thermodynamic cycle analysis, International Journal of Hydrogen Energy 36 (2011) 12141–12147.
[6] R. R. Bhosale, A. Kumar, P. Sutar, Thermodynamic analysis of solar driven SnO_2/SnO based thermochemical water splitting cycle, Energy Conversion & Management 135 (2017) 226-235.
[7] A. Stamatiou, P. Loutzenhiser, A. Steinfeld, Solar syngas production from H_2O and CO_2 via two-step thermochemical cycles based on Zn/ZnO and FeO/Fe_3O_4 redox reactions: Kinetic analysis, Energy & Fuels 24 (2010) 2716-2722.
[8] R.R. Bhosale, A. Kumar, F. AlMomani, Solar Thermochemical Hydrogen Production via Terbium Oxide Based Redox Reactions, International Journal of Photoenergy 2016 (2016) 9727895.
[9] P. Loutzenhiser, M. Gálvez, I. Hischier, A. Stamatiou, A. Frei, A. Steinfeld, CO_2 splitting via two-step solar thermochemical cycles with Zn/ZnO and FeO/Fe_3O_4 redox reactions II: Kinetic analysis, Energy & Fuels 23 (2009) 2832-2839.

[10] R. Bhosale, R. Shende, J. Puszynski, Sol-gel derived $NiFe_2O_4$ modified with ZrO_2 for hydrogen generation from solar thermochemical water-splitting reaction, Materials Research Society Symposium Proceedings 1387 (2011) 34 – 39.

[11] P. Charvin, S. Abanades, F. Lemont, G. Flamant, Experimental study of SnO2/SnO/Sn thermochemical systems for solar production of hydrogen, AIChE Journal 54 (2008) 2759-2767.

[12] R. V. Shende, J. Puszynski, M. Opoku, R. Bhosale, Synthesis of novel ferrite foam material for water-splitting application, Proceedings of the 2009 NSTI Nanotechnology Conference and Expo., 1 (2009) 201 – 204.

[13] C.C. Agrafiotis, C. Pagkoura, A. Zygogianni, G. Karagiannakis, M. Kostoglou, A.G. Konstandopoulos, Hydrogen production via solar-aided water splitting thermochemical cycles: Combustion synthesis and preliminary evaluation of spinel redox-pair materials, International Journal Hydrogen Energy 37 (2012) 8964-8980.

[14] R.R. Bhosale, A. Kumar, F. AlMomani, U. Ghosh, M. S. Anis, K. Kakosimos, R. V. Shende, M. A. Rosen, Solar hydrogen production via a samarium oxide-based thermochemical water splitting cycle, Energies 9 (2016).

[15] M. Roeb, N. Gathmann, M. Neises, C. Sattler, R. Pitz-Paal, Thermodynamic analysis of two-step solar water splitting with mixed iron oxides, International Journal of Energy Research 33 (2009) 893-902.

[16] R. R. Bhosale, A. Kumar, F. AlMomani, U. Ghosh, P. Sutar, G. Takalkar, A. Ashok, I. Alxneit, Effectiveness of Ni incorporation in iron oxide crystal structure towards thermochemical CO_2 splitting reaction, Ceramics International 43 (2017) 5150-5155.

[17] J.R. Scheffe, J. Li, A.W. Weimer, A spinel ferrite/hercynite water-splitting redox cycle, International Journal of Hydrogen Energy 35 (2010) 3333-3340.

[18] R. R. Bhosale, A. Kumar, F. AlMomani, U. Ghosh, D. Dardor, Z. Bouabidi, M. Ali, S. Yousefi, A. AlNouss, M. S. Anis, M. H. Usmani, M. H. Ali, R. S. Azzam, A. Banu, Solar co-production of samarium and syngas via methanothermal reduction of samarium sesquioxide, Energy Conversion and Management 112 (2016) 413 – 422.

[19] J.R. Scheffe, A. Steinfeld, Thermodynamic analysis of cerium-based oxides for solar thermochemical fuel production, Energy & Fuels 26 (2012) 1928-1936.

[20] R. R. Bhosale, R. V. Shende, J. Puszynski, Sol-gel synthesis of ferrite foam materials for H2 generation from water-splitting reaction, Proceedings of the 2010 NSTI Nanotechnology Conference and Expo., 1 (2010) 368 – 371.

[21] W.C. Chueh, C. Falter, M. Abbott, D. Scipio, P. Furler, S.M. Haile, A. Steinfeld, High-flux solar-driven thermochemical dissociation of CO_2 and H_2O using nonstoichiometric ceria, Science 330 (2010) 1797-1801.

[22] R. R. Bhosale, R. V. Shende, J. Puszynski, Synthesis of nano-sized ferrite materials for H_2 generation from high temperature water-splitting reaction, Proceedings of the 2011 NSTI Nanotechnology Conference and Expo., 1 (2010) 237 – 240.

[23] J.R. Scheffe, R. Jacot, G.R. Patzke, A. Steinfeld, Synthesis, Characterization, and Thermochemical Redox Performance of Hf+4 , Zr+4 , and Sc+3 Doped Ceria for Splitting CO2, The Journal of Physical Chemistry C 117 (2013) 24104-24114.

[24] R. R. Bhosale, R. V. Shende, J. Puszynski, H2 generation from thermochemical water-splitting using sol-gel synthesized ferrites (MxFeyOz, M = Ni, Zn, Sn, Co, Mn, Ce), Proceedings of the 2010 AIChE Annual Meeting (2010).

[25] J.R. Scheffe, D. Weibel, A. Steinfeld, Lanthanum–strontium–manganese perovskites as redox materials for solar thermochemical splitting of H_2O and CO_2, Energy & Fuels 27 (2013) 4250-4257.

[26] R. R. Bhosale, R. V. Shende, M. K. Opoku, J. Puszynski, H2 generation from water-splitting using novel ferrite foam materials, Proceedings of the 2009 AIChE Annual Meeting (2009).

[27] A.H. McDaniel, E.C. Miller, D. Arifin, A. Ambrosini, E.N. Coker, R. O'Hayre, W.C. Chueh, J. Tong, Sr-and Mn-doped $LaAlO_{3-\delta}$ for solar thermochemical H_2 and CO production, Energy & Environmental Science 6 (2013) 2424-2428.

[28] R. R. Bhosale, A. Kumar, A. Ashok, P. Sutar, G. D. Takalkar, M. Khraisheh, F. AlMomani, U. Ghosh, La-Based Perovskites as Oxygen-Exchange Redox Materials for Solar Syngas Production, MRS Advances 2 (2017) 3365-3370.

[29] M. Gálvez, R. Jacot, J. Scheffe, T. Cooper, G. Patzke, A. Steinfeld, Physico-chemical changes in Ca, Sr and Al-doped La–Mn–O perovskites upon thermochemical splitting of CO_2 via redox cycling, Physical Chemistry Chemical Physics 17 (2015) 6629-6634.

[30] R. R. Bhosale, P. Sutar, A. Kumar, F. AlMomani, M. H. Ali, U. Ghosh, S. Al-Muhtaseb, M. Khraisheh, Journal of Renewable and Sustainable Energy 8 (2016) 034702..

[31] F. A. Costa Oliveira, M. A. Barreiros, S. Abanades, A. P. F. Caetano, R. M. Novais, R. C. Pullar, Solar thermochemical CO2 splitting using cork-templated ceria ecoceramics, Journal of CO_2 Utilization 26 (2018) 552-563.

[32] R.R. Bhosale, A. Kumar, F. AlMomani, U. Ghosh, S. Al-Muhtaseb, R. Gupta, I. Alxneit, Assessment of $Ce_xZr_yHf_zO_2$ based oxides as potential solar thermochemical CO_2 splitting materials, Ceramics International 42 (2016) 9354-9362.

[33] M. Hoes, C. Muhich, R. Jacot, G. R. Patcke, A. Steinfeld, Thermodynamics of paired charge-compensating doped ceria with superior redox performance for solar thermochemical splitting of H_2O and CO_2, Journal of Material Chemistry A 5 (2017) 19476-19484.

[34] R.R. Bhosale, A. Kumar, F. AlMomani, I. Alxneit, Sol–gel derived CeO_2–Fe_2O_3 nanoparticles: Synthesis, characterization and solar thermochemical application, Ceramics International 42 (2016) 6728-6737.

[35] C. N. R. Rao, S. Dey, Generation of H_2 and CO by solar thermochemical splitting of H_2O and CO_2 by employing metal oxides, Journal of Solid State Chemistry 242 (2016) 107-115.

[36] R. R. Bhosale. A. Kumar, F. AlMomani, I. Alxneit, Propylene oxide assisted sol–gel synthesis of zinc ferrite nanoparticles for solar fuel production, Ceramics International 42 (2016) 2431-2438.

[37] Y. Hao, C. Yang, S. Haile, Ceria–Zirconia Solid Solutions ($Ce_{1-x}Zr_xO_{2-\delta}$, $x \leq 0.2$) for Solar Thermochemical Water Splitting: A Thermodynamic Study, Chemistry of Materials 26 (2014) 6073-6082.
[38] R. R. Bhosale. A. Kumar, F. AlMomani, U. Ghosh, M. Khraisheh, A comparative thermodynamic analysis of samarium and erbium oxide based solar thermochemical water splitting cycle, International Journal of Hydrogen Energy 42 (2017) 23416-23426.
[39] T. Cooper. J. Scheffe, M. E. Galvex, R. Jacot, G. Patzke, A. Steinfeld, Lanthanum Manganite Perovskites with Ca/Sr A-site and Al B-site Doping as Effective Oxygen Exchange Materials for Solar Thermochemical Fuel Production, Energy Technology 3 (2015) 1130 – 1142.
[40] R. R. Bhosale. Thermodynamic analysis of Ni-ferrite based solar thermochemical H2O splitting cycle for H_2 production, International Journal of Hydrogen Energy (2018) Article In-Press.
[41] C. Muhich, A. Steinfeld, Principles of doping ceria for the solar thermochemical redox splitting of H_2O and CO_2, Journal of Materials Chemistry A 5 (2017) 15578-15590.
[42] R. R. Bhosale, G. D. Takalkar, Nanostructured co-precipitated $Ce_{0.9}Ln_{0.1}O_2$ (Ln = La, Pr, Sm, Nd, Gd, Tb, Dy, or Er) for thermochemical conversion of CO_2, Ceramics International 44 (2018) 16688–16697.
[43] D. Arifin, A. W. Weimer, Kinetics and mechanism of solar-thermochemical H_2 and CO production by oxidation of reduced CeO_2, Solar Energy 160 (2018) 178-185.
[44] R. R. Bhosale, Synthesis and characterization of nanocrystalline $CoFe_2O_4$ – zirconia via propylene oxide aided sol-gel method, Ceramics International 44 (2018) 191-197.
[45] R. R. Bhosale, Thermodynamic efficiency analysis of zinc oxide based solar driven thermochemical H_2O splitting cycle: Effect of partial pressure of O_2, thermal reduction and H_2O splitting temperatures, International Journal of Hydrogen Energy 43 (2018) 14915–14924.
[46] R. R. Bhosale, Solar driven two-step CH_4 reforming and H_2O splitting using Al_2O_3 for Co-production of Al, syngas, and H_2, Solar Energy (2018) Article In-Press.

540768 Solid-Liquid and Vapor-Liquid Equilibria of BTEX Compounds in Methane and Ethane Mixtures at LNG Conditions

Abstract:

The formation and deposition of solids during the cryogenic processing of natural gas is a perennial risk for operators. If present in high enough quantities, heavy hydrocarbons including BTEX (benzene, cyclohexane, toluene, p-xylene and ethyl-benzene) aromatics and paraffins (n-hexane to n-decane) can lead to freeze-out problems in the cryogenic heat exchanger. A lack of experimental data for the solubility of such heavy hydrocarbons in LNG limits the accuracy of predictive engineering models used to avoid such problems. Moreover, data for the partitioning (VLE) of BTEX compounds, which exist as minor components in both phases, are particularly important given the freeze-out risk they pose for the downstream main cryogenic heat exchanger. Thus, in addition to the SLE data, high-quality VLE data for LNG-related mixtures are essential for improving and examining the accuracy of existing equations of state (EOS) used in the design and simulation of the cryogenic distillation columns, also known as scrub columns.

In this work, a specialized apparatus designed for visual measurements of solid-liquid equilibrium (SLE) was used to measure liquidus (melting) temperatures in binary mixtures of methane/ethane + p-xylene and methane + n-decane at varying compositions and pressures up to 30 MPa. Solid-liquid equilibrium data for p-xylene + methane were measured for the first time at compositions ranging from $0 < x_{C1} < 0.52$ at pressures to 28 MPa. Melting conditions for p- xylene + ethane mixtures were measured at compositions from $0 < x_{C2} < 0.95$ at temperatures down to 200.8 K. Predictions made using the Peng-Robinson EOS as implemented in industry-standard software under-estimated the melting temperatures of methane + p-xylene by over 8 K while for ethane + p-xylene, the melting temperatures were over-predicted by over 30 K. Melting temperatures of methane + decane systems were also measured down to 233.7 K at xC1 = 0.834 and p = 23 MPa.

Furthermore, an analytical apparatus, fitted with a GC for compositional analysis, was used to measure the vapor-liquid equilibrium of a ternary methane + propane + methylbenzene (toluene) and methane + propane + 1,4-dimethylbenzene (p-xylene) mixtures over a wide range of conditions, with toluene and p-xylene as the minor components in both the liquid and vapor phases. Measurements were conducted at temperatures between (213 and 323 K) and at pressures up to 12 MPa. The measured VLE data were compared to values calculated with the Groupe European de Recherche Gaziere (GERG-2008) multiparameter EOS and the HYSYS Peng Robinson (PR) equation of state (EOS) used widely in the LNG industry. The amount of toluene and p-xylene in the vapor phase was under-predicted by both EOS, with the deviations increasing in magnitude at low temperatures. These VLE measurements demonstrate that current EOS substantially under-predict the possible BTEX content of saturated vapors that could be present in the overhead product stream of an LNG scrub column.

542620 The Impact of Shale Gas and Oil on the Chemical Industry

Abstract:

Lower alkanes and aromatics are very useful as both fuels and chemical raw materials. Occasionally supply disruptions constrained availability but also encouraged investigation of alternative feedstocks and chemistries. Recent new technologies for the production of unconventional gas and oil sources from tight shale formations significantly increased availability of both methane and condensate, radically altered process economics, resulted in the construction of much new intermediates capacity, and multiplied the ways traditional and alternative chemistries might be exploited. This presentation will outline historically how natural gas components became useful as chemical feedstocks, how government regulations occasionally led to perturbations in availability and price, how supply crises led to the development of alternative chemistries, catalysts, and ultimately the exploitation of shale resources. More specifically, will be discussed how excess coproducts, especially condensate, from increasing shale gas and oil production for energy are having significant unexpected impacts on the chemical process industries and opportunities for new process chemistry research.

541186 Sustainable Energy and Chemicals: Past, Present, and Future

Abstract:

The last 40 years has seen the U.S. energy situation vacillate from gasoline rationing and concerns over Peak Oil, to projected energy and feedstock prices that were noncompetitive in global markets for chemicals production, to the economic revolution of frack gas and light tight oil. The importance of low cost energy and feedstocks in driving economic opportunity will be discussed relative to this history, along with the importance of sustainability in decision making to gain stakeholder buy-in and acceptance. Strategies for managing career paths and competencies despite changing targets and high uncertainty will be addressed.

542618 Disruptions: What the Future May Hold

Abstract:

Disruptive forces are all around us. They change the way we communicate and work. They change the way we develop goods and services. They also change the types of products that are manufactured. Examples such as the way in which we produce and consume energy or the feedstocks we use to produce product have seen many changes over the course of the last few decades, and these disruptions are expected to continue. The chemical engineering discipline and the expectations for a chemical engineerâ€™s skills has also changed and adapted as a result of these disruptive forces. A discussion of the broad themes associated with these forces and the implications that they may have to chemical engineers in the future will be discussed.

542621 Geopolitical Factors Influencing the Evolution of the Chemical Industry

Abstract:

The basic chemicals and plastics industry is built on a simple principle: find a cheap feedstock and convert it into something more valuable. However, the right feedstock depends on place and time. Certain policies and actions of various governments have influenced the evolution and profitability of the chemical industry during the past 50 years, sometimes in surprising ways. For example, high oil prices and US government funding encouraged the development and commercialization of coal gasification technology during the late 1970s and early 1980s. This enabled China to develop a competitive basic chemical industry using coal as a feedstock during the last 25 years. Similarly, an abrupt increase in oil prices during the first decade of the 21st century encouraged the present shale gas revolution, causing a major shift in the economy of producing basic chemicals and polymers in the US. If history is our guide, then what governments choose to fund and regulate may be as important as market forces in determining the shape of our industry and profession in the future.

542622 Agility & Resilience: How to Maintain Career Competitiveness in the Changing Chemical Industry

Abstract:

The world around is changing. The shift to protectionism, global instability and price volatility of raw materials and feedstocks has the potential to wreak havoc on the chemical industry. With the revolution of shale gas in North America and the sudden drop in feedstock costs, the chemical sector in the US has reasserted its historic dominance. Learn how the role of a chemical engineer can contribute to industry growth and resilience, while adapting to allow for personal agility throughout the career life cycle.

542626 PSA Based Warm Gas CO_2 Capture from Synthesis Gas

Abstract:

TDA is developing a novel sorbent based PSA process that removes CO_2 via physical adsorption from synthesis gas. The relatively strong affinity of the sorbent to CO_2 enables effective operation at temperatures up to 300°C (well above the dew point of synthesis gas stream generated by commercial gasifiers). However, because the sorbent and the CO_2 do not form a true covalent bond, the energy needed to regenerate TDAâ€™s sorbent (5.4 kcal per mol of CO_2) is much lower than that observed for either chemical absorbents (e.g., 29.9 kcal/mol CO_2 for sodium carbonate) or amine-based solvents (e.g., 14.2 kcal/mol CO_2 for monoethanolamine). TDAâ€™s sorbent can be regenerated isothermally and CO_2 can be recovered at pressure (as high as 150 psia). Thus, the energy needed to regenerate the sorbent and compress the CO_2 for sequestration is significantly lower than that for any other technology reported to date. The high surface area and favorable porosity of the sorbent also provides a unique platform to introduce additional functionality, such as active groups to catalyze the water-gas-shift (WGS) reaction or to remove trace metals (e.g., Hg, As). Hence, TDAâ€™s CO_2 sorbent could be used to do a warm gas capture of CO_2 above the dew point of the synthesis gas in an Integrated Gasification Combined Cycle (IGCC) power plant. TDA's pre-combustion CO_2 capture system achieves a net plant efficiency of 34% on a coal high heating value basis when operated on bituminous coal, which is significantly higher than that can be achieved with the Selexol™ scrubber at 31.6% (state-of-the-art pre-combustion carbon capture technology).

In a current DOE sponsored project, TDA is carrying out field tests with the warm gas CO_2 capture unit integrated with WGS catalyst at 10 cfm scale. In this paper we present the results from the 8-bed PSA cycle performance in the field test with real synthesis gas.

Solar Thermochemical CO_2 Splitting using $Ce_xZn_{1-x}O_2$ Derived via Co-precipitation Method

(Principal Author)
Gorakshnath Takalkar
Department of Chemical Engineering
College of Engineering
Qatar University, Doha, Qatar

(Co-Authors)
Rahul Bhosale
Department of Chemical Engineering
College of Engineering
Qatar University, Doha, Qatar
rahul.bhosale@qu.edu.qa

Aliya Banu
Department of Chemical Engineering
College of Engineering
Qatar University, Doha, Qatar

Fares AlMomani
Department of Chemical Engineering
College of Engineering
Qatar University, Doha, Qatar

Anand Kumar
Department of Chemical Engineering
College of Engineering
Qatar University, Doha, Qatar

Majeda Khraisheh
Department of Chemical Engineering
College of Engineering
Qatar University, Doha, Qatar

Prepared for Presentation at
American Institute of Chemical Engineers
2018 Fall Meeting
Pittsburgh, PA, USA
October 28 – November 2, 2018

Solar Thermochemical CO_2 Splitting using $Ce_xZn_{1-x}O_2$ Derived via Co-precipitation Method

Gorakshnath Takalkar, Rahul R. Bhosale*, Aliya Banu, Fares AlMomani, Anand Kumar, Majeda Khraisheh

Department of Chemical Engineering
College of Engineering
Qatar University, Doha, Qatar

Keywords: Ceria, Zinc Oxide, Thermochemical, CO_2 splitting, Co-Precipitation

Abstract

This study reports the synthesis of the Zn doped ceria redox materials using the co-precipitation of hydroxide method for the application towards thermochemical CO_2 splitting reactions. The thermochemical CO_2 splitting ability of the derived materials was also identified by conducting thermogravimetric experiments.

1. Introduction

Ceria based redox materials are considered as the promising option available from a variety of metal oxides investigated for solar thermochemical splitting of H_2O and CO_2[1-46]. In our recent study, we have synthesized transition metal doped ceria ($Ce_{0.9}M_{0.1}O_2$, M = Ni, Zn, Mn, Fe, Cu, Cr, Co, Zr) by using co-precipitation of hydroxide method and tested these materials towards thermochemical CO_2 splitting. The obtained results indicate that the Zn doped ceria shows promising characteristics towards both thermal reduction and CO_2 splitting steps. However, it is important to note that only 10% of the Zn was doped in the ceria crystal structure. There is a strong chance that if the percentage of doping is varied, the thermal reduction and CO_2 splitting capacity of the Ceria-Zinc oxide can be enhanced further. Therefore, in this study, we have synthesized $Ce_xZn_{1-x}O_2$ redox materials by using the same synthesis approach and tested them towards solar thermochemical CO_2 splitting reaction using a thermogravimetric analyzer. The reaction chemistry associated with the $Ce_xZn_{1-x}O_2$ based thermochemical CO_2 splitting is shown below:

$$Ce_xZn_{1-x}O_2 \rightarrow Ce_xZn_{1-x}O_{2-\delta} + O_2(g) \quad (1)$$

$$Ce_xZn_{1-x}O_{2-\delta} + CO_2 \rightarrow Ce_xZn_{1-x}O_2 + CO(g) \quad (2)$$

2. Experimental

2.1 Materials

To synthesize $Ce_xZn_{1-x}O_2$ redox materials, the nitrate precursors of Ce and Zn were purchased from Sigma Aldrich. Likewise, the precipitating agent, aqueous ammonia hydroxide was procured from Sigma Aldrich. All the samples were prepared using the deionized water produced from ultrapure Type 1 from Direct –Q system, Millipore, France. The gases utilized in the TGA experiments are acquired from Buzwair Scientific and Technical Gases supplier, Qatar.

2.2 Co-precipitation synthesis of $Ce_xZn_{1-x}O_2$

To achieve a successful preparation of $Ce_xZn_{1-x}O_2$ redox materials, exactly calculated amounts of the metal precursors were dissolved in the deionized water. To this solution, aqueous ammonia hydroxide was added in a dropwise manner to maintain the solution pH approximately equal to 10. Stirring of the aqueous dispersion is carried out for 48 hours (with a maintained pH ~ 10). After the 2 days of continuous stirring, the solution obtained was allowed to settle down overnight. During this settling time, the precipitate will settle down and this precipitate was washed for multiple times using deionized water (by using vacuum filtration kit). The solids collected after filtration were dried, crushed and annealed in presence of air up to 800 °C for 4 hours. The obtained calcined powder was stored and then analyzed further towards characterization and TGA experiments. During the synthesis of $Ce_xZn_{1-x}O_2$ redox materials via co-precipitation of hydroxide, the change in the color of the solution is presented in Figure 1.

2.3 CO_2 Splitting Experiments

The $Ce_xZn_{1-x}O_2$ redox materials were examined towards thermochemical CO_2 splitting experiments using a thermogravimetric analyzer show in Figure 2. The temperatures employed for this study are selected based on our previously published paper (associated with the $Ce_{0.9}M_{0.1}O_2$). The powder to be tested underwent multiple thermochemical thermal reduction and CO_2 splitting steps to identify the O_2 release and CO production capacity. The loss and gain in the mass of the sample can be directly converted into the amounts of O_2 released and CO produced. The buoyancy effect was eliminated by subtracting the blank run from the actual experiments.

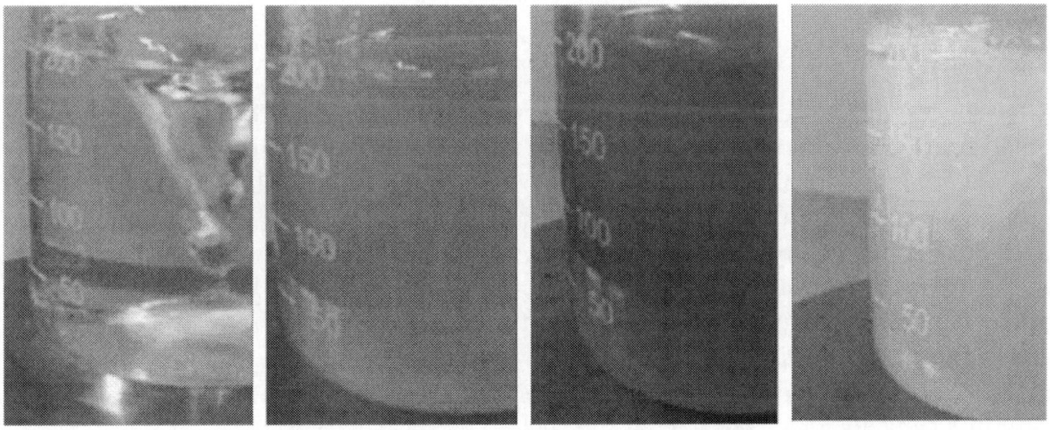

Figure 1. Variation in the color of the precipitating solution during synthesis of $Ce_xZn_{1-x}O_2$.

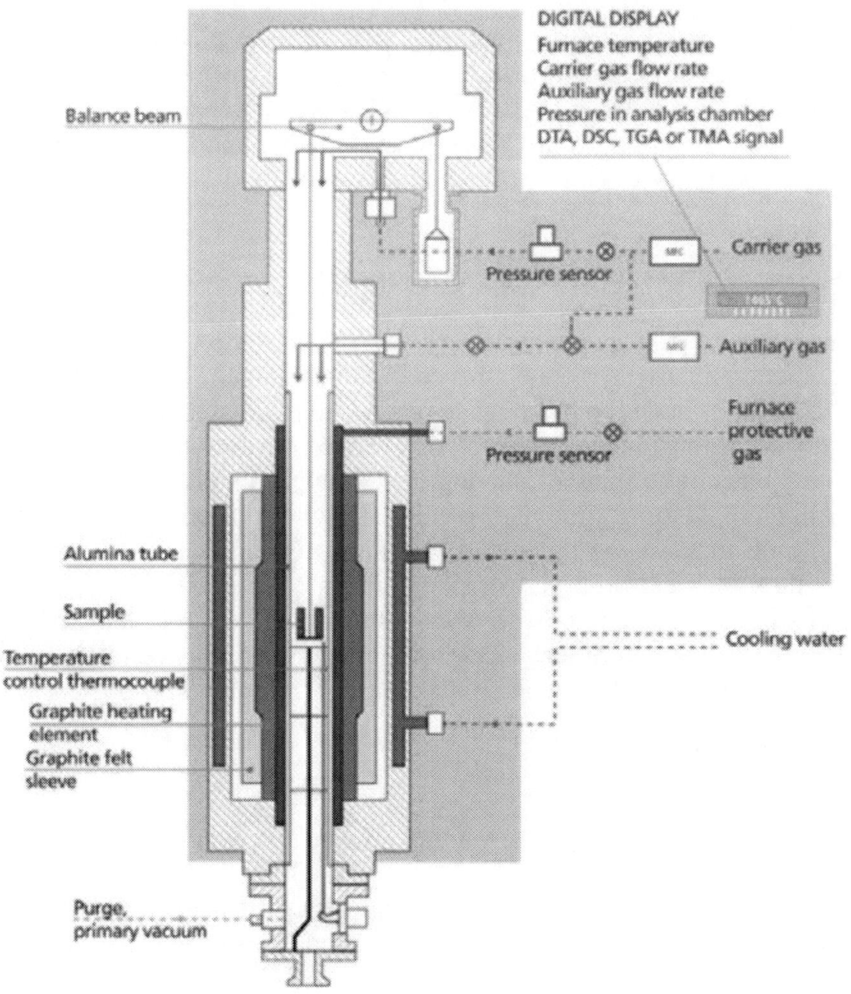

Figure 2. Thermogravimetric analyzer used for the splitting of CO_2 using $Ce_xZn_{1-x}O_2$.

3. Results and Discussion

After successfully synthesizing and characterizing the $Ce_xZn_{1-x}O_2$ redox materials, we have attempted few thermogravimetric CO_2 splitting experiments to identify the redox potential of the derived materials. According to the comparison performed on the basis of 1st thermochemical cycle, the obtained results indicate that as the Zn concentration in the ceria increases, the amounts of O_2 released enhances. In contracts, the same increase in the molar concentration of Zn in the ceria crystal structure impedes the CO_2 splitting ability of the $Ce_xZn_{1-x}O_2$ redox materials. However, these results are only of indicative manner. We are currently working towards performing multiple thermochemical cycles using all the derived $Ce_xZn_{1-x}O_2$ redox materials to determine the best possible combination in terms of better thermal reduction and CO_2 splitting capacity.

4. Conclusions

By employing the co-precipitation of hydroxide method, several combinations of the $Ce_xZn_{1-x}O_2$ redox materials were synthesized and characterized successfully. The initial thermogravimetric thermochemical CO_2 splitting experiments indicate that the increase in the molar concentration of Zn in the ceria crystal structure improves the thermal reduction capacity. However, the CO_2 splitting ability was observed to be decreased due to the similar rise in the Zn molar concentration.

5. Acknowledgments

This publication was made possible by the NPRP grant (NPRP8-370-2-154) from the Qatar National Research Fund (a member of Qatar Foundation). The statements made herein are solely the responsibility of author(s).

References

[1] J.R. Scheffe, A. Steinfeld, Oxygen exchange materials for solar thermochemical splitting of H_2O and CO_2: a review, Materials Today 17 (2014) 341–348.
[2] R. R. Bhosale, G. D. Takalkar, P. N. Sutar, A. Kumar, F. AlMomani, M. Khraisheh, A decade of ceria based solar thermochemical H_2O/CO_2 splitting cycle, International Journal of Hydrogen Energy (2018) Article In-Press.
[3] D. Yadav, R. Banerjee, A review of solar thermochemical processes, Renewable and Sustainable Energy Reviews 54 (2016) 497 – 532.[4] R. R. Bhosale, A. Kumar, F. AlMomani, R. B. Gupta, Solar thermochemical $ZnO/ZnSO_4$ water splitting cycle for hydrogen production, International Journal of Hydrogen Energy 43 (2017) 23474-23483.
[5] P.G. Loutzenhiser, A. Steinfeld, Solar syngas production from CO2 and H2O in a two-step thermochemical cycle via Zn/ZnO redox reactions: thermodynamic cycle analysis, International Journal of Hydrogen Energy 36 (2011) 12141–12147.
[6] R. R. Bhosale, A. Kumar, P. Sutar, Thermodynamic analysis of solar driven SnO_2/SnO based thermochemical water splitting cycle, Energy Conversion & Management 135 (2017) 226-235.
[7] A. Stamatiou, P. Loutzenhiser, A. Steinfeld, Solar syngas production from H_2O and CO_2 via two-step thermochemical cycles based on Zn/ZnO and FeO/Fe_3O_4 redox reactions: Kinetic analysis, Energy & Fuels 24 (2010) 2716-2722.
[8] R.R. Bhosale, A. Kumar, F. AlMomani, Solar Thermochemical Hydrogen Production via Terbium Oxide Based Redox Reactions, International Journal of Photoenergy 2016 (2016) 9727895.
[9] P. Loutzenhiser, M. Gálvez, I. Hischier, A. Stamatiou, A. Frei, A. Steinfeld, CO_2 splitting via two-step solar thermochemical cycles with Zn/ZnO and FeO/Fe_3O_4 redox reactions II: Kinetic analysis, Energy & Fuels 23 (2009) 2832-2839.

[10] R. Bhosale, R. Shende, J. Puszynski, Sol-gel derived $NiFe_2O_4$ modified with ZrO_2 for hydrogen generation from solar thermochemical water-splitting reaction, Materials Research Society Symposium Proceedings 1387 (2011) 34 – 39.

[11] P. Charvin, S. Abanades, F. Lemont, G. Flamant, Experimental study of SnO2/SnO/Sn thermochemical systems for solar production of hydrogen, AIChE Journal 54 (2008) 2759-2767.

[12] R. V. Shende, J. Puszynski, M. Opoku, R. Bhosale, Synthesis of novel ferrite foam material for water-splitting application, Proceedings of the 2009 NSTI Nanotechnology Conference and Expo., 1 (2009) 201 – 204.

[13] C.C. Agrafiotis, C. Pagkoura, A. Zygogianni, G. Karagiannakis, M. Kostoglou, A.G. Konstandopoulos, Hydrogen production via solar-aided water splitting thermochemical cycles: Combustion synthesis and preliminary evaluation of spinel redox-pair materials, International Journal Hydrogen Energy 37 (2012) 8964-8980.

[14] R.R. Bhosale, A. Kumar, F. AlMomani, U. Ghosh, M. S. Anis, K. Kakosimos, R. V. Shende, M. A. Rosen, Solar hydrogen production via a samarium oxide-based thermochemical water splitting cycle, Energies 9 (2016).

[15] M. Roeb, N. Gathmann, M. Neises, C. Sattler, R. Pitz-Paal, Thermodynamic analysis of two-step solar water splitting with mixed iron oxides, International Journal of Energy Research 33 (2009) 893-902.

[16] R. R. Bhosale, A. Kumar, F. AlMomani, U. Ghosh, P. Sutar, G. Takalkar, A. Ashok, I. Alxneit, Effectiveness of Ni incorporation in iron oxide crystal structure towards thermochemical CO_2 splitting reaction, Ceramics International 43 (2017) 5150-5155.

[17] J.R. Scheffe, J. Li, A.W. Weimer, A spinel ferrite/hercynite water-splitting redox cycle, International Journal of Hydrogen Energy 35 (2010) 3333-3340.

[18] R. R. Bhosale, A. Kumar, F. AlMomani, U. Ghosh, D. Dardor, Z. Bouabidi, M. Ali, S. Yousefi, A. AlNouss, M. S. Anis, M. H. Usmani, M. H. Ali, R. S. Azzam, A. Banu, Solar co-production of samarium and syngas via methanothermal reduction of samarium sesquioxide, Energy Conversion and Management 112 (2016) 413 – 422.

[19] J.R. Scheffe, A. Steinfeld, Thermodynamic analysis of cerium-based oxides for solar thermochemical fuel production, Energy & Fuels 26 (2012) 1928-1936.

[20] R. R. Bhosale, R. V. Shende, J. Puszynski, Sol-gel synthesis of ferrite foam materials for H2 generation from water-splitting reaction, Proceedings of the 2010 NSTI Nanotechnology Conference and Expo., 1 (2010) 368 – 371.

[21] W.C. Chueh, C. Falter, M. Abbott, D. Scipio, P. Furler, S.M. Haile, A. Steinfeld, High-flux solar-driven thermochemical dissociation of CO_2 and H_2O using nonstoichiometric ceria, Science 330 (2010) 1797-1801.

[22] R. R. Bhosale, R. V. Shende, J. Puszynski, Synthesis of nano-sized ferrite materials for H_2 generation from high temperature water-splitting reaction, Proceedings of the 2011 NSTI Nanotechnology Conference and Expo., 1 (2010) 237 – 240.

[23] J.R. Scheffe, R. Jacot, G.R. Patzke, A. Steinfeld, Synthesis, Characterization, and Thermochemical Redox Performance of Hf+4 , Zr+4 , and Sc+3 Doped Ceria for Splitting CO2, The Journal of Physical Chemistry C 117 (2013) 24104-24114.

[24] R. R. Bhosale, R. V. Shende, J. Puszynski, H2 generation from thermochemical water-splitting using sol-gel synthesized ferrites (MxFeyOz, M = Ni, Zn, Sn, Co, Mn, Ce), Proceedings of the 2010 AIChE Annual Meeting (2010).

[25] J.R. Scheffe, D. Weibel, A. Steinfeld, Lanthanum–strontium–manganese perovskites as redox materials for solar thermochemical splitting of H_2O and CO_2, Energy & Fuels 27 (2013) 4250-4257.

[26] R. R. Bhosale, R. V. Shende, M. K. Opoku, J. Puszynski, H2 generation from water-splitting using novel ferrite foam materials, Proceedings of the 2009 AIChE Annual Meeting (2009).

[27] A.H. McDaniel, E.C. Miller, D. Arifin, A. Ambrosini, E.N. Coker, R. O'Hayre, W.C. Chueh, J. Tong, Sr-and Mn-doped $LaAlO_{3-\delta}$ for solar thermochemical H_2 and CO production, Energy & Environmental Science 6 (2013) 2424-2428.

[28] R. R. Bhosale, A. Kumar, A. Ashok, P. Sutar, G. D. Takalkar, M. Khraisheh, F. AlMomani, U. Ghosh, La-Based Perovskites as Oxygen-Exchange Redox Materials for Solar Syngas Production, MRS Advances 2 (2017) 3365-3370.

[29] M. Gálvez, R. Jacot, J. Scheffe, T. Cooper, G. Patzke, A. Steinfeld, Physico-chemical changes in Ca, Sr and Al-doped La–Mn–O perovskites upon thermochemical splitting of CO_2 via redox cycling, Physical Chemistry Chemical Physics 17 (2015) 6629-6634.

[30] R. R. Bhosale, P. Sutar, A. Kumar, F. AlMomani, M. H. Ali, U. Ghosh, S. Al-Muhtaseb, M. Khraisheh, Journal of Renewable and Sustainable Energy 8 (2016) 034702..

[31] F. A. Costa Oliveira, M. A. Barreiros, S. Abanades, A. P. F. Caetano, R. M. Novais, R. C. Pullar, Solar thermochemical CO2 splitting using cork-templated ceria ecoceramics, Journal of CO_2 Utilization 26 (2018) 552-563.

[32] R.R. Bhosale, A. Kumar, F. AlMomani, U. Ghosh, S. Al-Muhtaseb, R. Gupta, I. Alxneit, Assessment of $Ce_xZr_yHf_zO_2$ based oxides as potential solar thermochemical CO_2 splitting materials, Ceramics International 42 (2016) 9354-9362.

[33] M. Hoes, C. Muhich, R. Jacot, G. R. Patcke, A. Steinfeld, Thermodynamics of paired charge-compensating doped ceria with superior redox performance for solar thermochemical splitting of H_2O and CO_2, Journal of Material Chemistry A 5 (2017) 19476-19484.

[34] R.R. Bhosale, A. Kumar, F. AlMomani, I. Alxneit, Sol–gel derived CeO_2–Fe_2O_3 nanoparticles: Synthesis, characterization and solar thermochemical application, Ceramics International 42 (2016) 6728-6737.

[35] C. N. R. Rao, S. Dey, Generation of H_2 and CO by solar thermochemical splitting of H_2O and CO_2 by employing metal oxides, Journal of Solid State Chemistry 242 (2016) 107-115.

[36] R. R. Bhosale. A. Kumar, F. AlMomani, I. Alxneit, Propylene oxide assisted sol–gel synthesis of zinc ferrite nanoparticles for solar fuel production, Ceramics International 42 (2016) 2431-2438.

[37] Y. Hao, C. Yang, S. Haile, Ceria–Zirconia Solid Solutions ($Ce_{1-x}Zr_xO_{2-\delta}$, $x \leq 0.2$) for Solar Thermochemical Water Splitting: A Thermodynamic Study, Chemistry of Materials 26 (2014) 6073-6082.
[38] R. R. Bhosale. A. Kumar, F. AlMomani, U. Ghosh, M. Khraisheh, A comparative thermodynamic analysis of samarium and erbium oxide based solar thermochemical water splitting cycle, International Journal of Hydrogen Energy 42 (2017) 23416-23426.
[39] T. Cooper. J. Scheffe, M. E. Galvex, R. Jacot, G. Patzke, A. Steinfeld, Lanthanum Manganite Perovskites with Ca/Sr A-site and Al B-site Doping as Effective Oxygen Exchange Materials for Solar Thermochemical Fuel Production, Energy Technology 3 (2015) 1130 – 1142.
[40] R. R. Bhosale. Thermodynamic analysis of Ni-ferrite based solar thermochemical H2O splitting cycle for H_2 production, International Journal of Hydrogen Energy (2018) Article In-Press.
[41] C. Muhich, A. Steinfeld, Principles of doping ceria for the solar thermochemical redox splitting of H_2O and CO_2, Journal of Materials Chemistry A 5 (2017) 15578-15590.
[42] R. R. Bhosale, G. D. Takalkar, Nanostructured co-precipitated $Ce_{0.9}Ln_{0.1}O_2$ (Ln = La, Pr, Sm, Nd, Gd, Tb, Dy, or Er) for thermochemical conversion of CO_2, Ceramics International 44 (2018) 16688–16697.
[43] D. Arifin, A. W. Weimer, Kinetics and mechanism of solar-thermochemical H_2 and CO production by oxidation of reduced CeO_2, Solar Energy 160 (2018) 178-185.
[44] R. R. Bhosale, Synthesis and characterization of nanocrystalline $CoFe_2O_4$ – zirconia via propylene oxide aided sol-gel method, Ceramics International 44 (2018) 191-197.
[45] R. R. Bhosale, Thermodynamic efficiency analysis of zinc oxide based solar driven thermochemical H_2O splitting cycle: Effect of partial pressure of O_2, thermal reduction and H_2O splitting temperatures, International Journal of Hydrogen Energy 43 (2018) 14915–14924.
[46] R. R. Bhosale, Solar driven two-step CH_4 reforming and H_2O splitting using Al_2O_3 for Co-production of Al, syngas, and H_2, Solar Energy (2018) Article In-Press.
[47] G.D. Takalkar, R.R. Bhosale, A. Kumar, F. AlMomani, M. Khraisheh, R.A. Shakoor, R.B. Gupta, Transition metal doped ceria for solar thermochemical fuel production, Solar Energy (2018) Article In-Press.

525156 Successful Demonstration of Illinois Industrial Carbon Capture and Storage in a Saline Reservoir

Abstract:

This paper presents an overview and initial operating results of the Archer Daniels Midland Company's (ADM) Illinois Industrial Carbon Capture and Storage (ICCS) project in Decatur, Illinois. In collaboration with the U.S. Department of Energy (DOE), ADM has completed construction of the compression and storage facilities and initiated carbon dioxide storage (sequestration) in 2017. As of March 2018, this project has safely and permanently stored approximately 640,000 metric tons of carbon dioxide deep underground.

This project has been demonstrating an integrated system for capturing carbon dioxide from an ethanol production plant and geologically storing it on site, in the Mount Simon Sandstone, a saline reservoir, which covers portions of the Midwest including central and southern Illinois. The carbon dioxide produced is a byproduct of processing corn into fuel-grade ethanol at the ADM plant. The Illinois ICCS is the largest saline storage project in the United States. The Office of Fossil Energy's National Energy Technology Laboratory (NETL) manages this project, which received $141.4 million in DOE funding and another $66.5 million in private sector cost-sharing.

This is the first project to operate with the U.S. Environmental Protection Agency's Class VI injection well permit for geological storage of carbon dioxide. This project is demonstrating cutting-edge technologies for intelligent monitoring. The lessons learned during the initial operations will be presented. Because of excellent site geological conditions, there were no significant impacts of carbon dioxide storage on the reservoir integrity, underground sources of drinking water, and induced seismicity. The impacts of carbon dioxide injection on the reservoir pressure and temperature, pressure propagation, carbon dioxide plume migration, and carbon dioxide concentrations in the sandstone formation fluids will be highlighted. Schlumberger's computational models were used to predict the reservoir response to carbon dioxide storage. These computational model predictions were compared with the actual data observed during the site monitoring and the results will be presented.

This project site has become a promising target for researchers interested in developing advanced engineering & computational models and analytical instruments for subsurface monitoring of carbon storage. For example, at this site, under a DOE cost shared project, ADM, in collaboration with Lawrence Berkeley National Laboratory and others, is demonstrating an intelligent monitoring system (IMS) for geologic storage and reservoir management. The status of this IMS project and its benefits to the ongoing Illinois ICCS project will be highlighted.

Successful implementation of this project could facilitate exploration of long-term carbon dioxide utilization options, such as enhanced oil recovery in the Southern Illinois Basin. Because ethanol plants produce a high purity carbon dioxide stream (i.e. low capture cost), these plants are low-hanging fruits for developing economical enhanced oil recovery projects (& related carbon dioxide pipeline infrastructure) and boosting domestic oil production to support the America First Energy Plan.

528359 Chemical Looping Combustion (CLC)-Aided Biomass Gasification for Co-Production of Hydrogen and Electricity

Abstract:

This paper presents a comprehensive thermodynamic evaluation of multiple carbon capture pathways of utilizing chemical looping combustion (CLC) to aid biomass gasification for co-production of electricity and fuels.

Gasification of carbonaceous fuels produces synthesis gas (syngas), consisting predominantly of H_2 and CO, which could be used as a feedstock for a wide range of chemical processes, including production of fuels such as hydrogen or Fischer-Tropsch (FT) liquids. Biomass gasification for production of syngas, in combination with carbon capture and storage (CCS), potentially leads to net negative CO_2 emissions. Biomass gasification technology, using oxygen, steam or CO_2 has developed over the years with a few large-scale demonstration plants being operated. The composition of syngas depends on the gasifying agent and operating conditions of the gasifier. Gasification using high-purity O_2, instead of air, produces a high heating value syngas. However, the cryogenic air separation unit (ASU) required to produce high-purity O_2 is an energy-intensive and a costly process. Gasification using steam also increases the syngas heating values by generating more hydrogen. However, the overall thermal conversion efficiency of steam gasification of biomass was reported to be in the range of 50 â€" 70% owing to the high demand of energy for producing steam and maintaining high gasification temperatures. On the other hand, CO_2 gasification has been reported to attain higher carbon and energy efficiency. Thermodynamic analysis has shown higher conversion of biomass gasification through recycling of CO_2.

Chemical looping combustion (CLC) is an indirect combustion process in which fuel is combusted without direct contact with air. Transfer of oxygen between air and fuel takes place with the aid of an oxygen-carrier (OC), usually oxides of transient metals. The reduced form of metal oxide extracts O_2 from air in one reactor and then transfers it to fuel in a subsequent reactor. Since the fuel does not come in direct contact with air, the products of combustion contain only carbon dioxide (CO_2) and water (H_2O). CO_2 stream of very high purity can be obtained by condensing the water vapor. Gaseous fuels such as natural gas or solid fuels have been demonstrated as fuels in CLC, for inherent CO_2 capture.

In this study, a novel approach in integrating biomass gasification, aided by CLC is investigated, with a view of attaining lower carbon emissions. The hot CO_2 and H_2O products from the fuel reactor of CLC are used as gasifying agents for biomass, while enabling auto-thermal operation and eliminating the need for expensive processes for O_2 or steam production. This will also greatly reduce the demand for process water required to generate steam. The hot depleted air stream from the air reactor is used to further generate steam.

Separate performance models are developed to perform a detailed thermodynamic analysis of:

(1) CLC reactor system using natural gas or syngas as fuel and Ni-/Cu-/Fe-based materials as oxygen carriers;

(2) biomass gasification using CO_2 and steam as gasifying agents; and

(3) combined CLC and biomass gasifier system.

The effect of varying operating conditions (temperature, steam-carbon and CO_2-carbon ratio, etc.) on performance metrics such as the syngas composition and heating value, overall thermal efficiency, and overall CO_2 emissions is studied. A comparison will be made with a conventional biomass gasification system in order to understand the relative feasibility of CLC-aided biomass gasification. The performance models are developed using Aspen Plus and MATLAB.

Using the results of the combined models, we will explore the technical feasibility of different utilization pathways of the syngas generated in the above process, specifically for co-production of H_2 and electricity in an integrated gasification combined cycle power plant (IGCC), with pre-combustion CO_2 capture and storage.

This work will help identify potential pathways to utilize CLC and biomass to achieve net-negative CO_2 emissions.

530790 Molecular Insights into the Enhanced Rate of CO_2 Absorption to Produce Bicarbonate in Aqueous 2-Amino-2-Methyl-1-Propanol (AMP)

Abstract:

Carbon dioxide (CO_2) capture from the flue gases of fossil fuel-fired power plants is needed to curb rising anthropogenic greenhouse gas emissions and mitigate global warming.[1] Wet scrubbing using aqueous amine-based solvents currently appears the most suitable method to remove CO_2 from flue gas because of their ability to chemically absorb CO_2 at low partial pressure and release the absorbed CO_2 with adequate kinetics.[2–4] Monoethanolamine (MEA) has been considered the benchmark solvent, but its widespread implementation remains hindered by high costs associated with the parasitic energy consumption during solvent regeneration in addition to degradation and corrosion issues.[5] Tertiary amines such as methyldiethanolamine (MDEA) may be more resistant to degradation and corrosion in addition to exhibiting higher CO_2 absorption capacities than MEA, but they suffer from slow absorption rates.[6,7] Sterically hindered amines, including 2-amino-2-methyl-1-propanol (AMP), typically have similar advantageous qualities as tertiary amines while exhibiting relatively much higher CO_2 absorption rates.[8–11] However, the molecular mechanisms underlying the enhanced absorption rate of sterically hindered amines relative to tertiary amines remains unclear.

Primary/secondary amines including MEA predominantly form carbamate, which is now widely accepted to occur via a two-step mechanism with a zwitterion intermediate; for instance, $2MEA + CO_2 \rightarrow MEA^+COO^- \rightarrow MEACOO^- + MEAH^+$, limiting the loading capacity to 0.5 mol CO_2/mol MEA.[12,13] Tertiary amines predominantly form bicarbonate (HCO_3^-), which may occur via base-catalyzed hydration of CO_2, e.g. $MDEA + H_2O + CO_2 \rightarrow MDEAH^+ + HCO_3^-$.[6,14] Experimental measurements show that CO_2 capture in tertiary amines tends to follow the Bronsted relationship where the logarithm of the absorption rate is linearly dependent on the pKa value.[15,16] However, sterically hindered amines typically exhibit much higher rates than expected from their basicities. For example, AMP (pKa = 9.70) has a similar basicity to diethylethanolamine (DEEA, pKa = 9.81), but the former (sterically hindered primary amine) forms bicarbonate an order of magnitude faster than the latter (tertiary amine).[7,17–19]

Due to this significant rate increase, it has been often speculated that AMP may first react with CO_2 to form carbamate, but the rather unstable carbamate due to the presence of bulky methyl groups may subsequently undergo hydrolysis to form bicarbonate, i.e., $AMPCOO^- + H_2O \rightarrow AMPH^+ + HCO_3^-$).[8,19] On the contrary, recent first-principles calculations predicted that AMP carbamate can be as stable as MEA carbamate in aqueous solution, and thus the former would not undergo hydrolysis more readily relative to the latter.[20] Very recently, through comparative theoretical study of AMP and MEA for CO_2 capture we have found that bicarbonate formation seems to be kinetically more probable in aqueous AMP while carbamate is more likely to form in aqueous MEA. It appears that the preferred bicarbonate formation is largely attributed to the relatively strong interaction between N (in AMP) and H (H_2O) which suppresses the reaction with CO_2 to form carbamate.[20]

This theoretical work focuses on elucidating the molecular mechanisms responsible for the enhanced rate of bicarbonate formation during CO_2 capture in aqueous AMP in comparison to DEEA.[21] Ab initio (Car-Parrinello) MD simulations in combination with metadynamics were employed to construct the free energy surfaces for CO_2 hydration catalyzed by AMP and DEEA. Our simulations unequivocally predict the free energy barrier to be lower in aqueous AMP (= 8.1 kcal mol^{-1}) compared to DEEA (= 11.4 kcal mol^{-1}). The amine-catalyzed hydration of CO_2 is found to require a lower barrier than that for its reaction with a free hydroxide anion (OH^-) in aqueous medium (=13.4 kcal mol^{-1}), which can be well explained by the stronger hydration of free OH^- in pure aqueous conditions. Analysis of MD simulation trajectories hints that the free energy barrier can largely be influenced by reorganization of H_2O molecules around the basic N site to stabilize the transition state involving OH^- formation and CO_2 polarization.

We further examined the entropic effects by evaluating the arrangement of H_2O molecules around AMP and DEEA in aqueous solution based on corresponding radial distribution functions (RDFs) calculated from AIMD simulations. The neighboring H_2O molecules around the N of DEEA (N_{DEEA}), relative to the AMP case, are more ordered and more hindered from facile rearrangement. The suppressed thermal displacement is largely attributed to the stronger hydrogen bonding of H_2O with N_{DEEA} than N_{AMP} in addition to the presence of bulky ethyl groups on N_{DEEA}. The hindered reorganization of neighboring H_2O molecules causes an increase in the free energy barrier for proton abstraction by DEEA compared to AMP due to a relatively lower stability of the transition state involving OH^- creation, and prevents the approach of CO_2 to the reaction region near the N_{DEEA} site, relative to N_{AMP}, as confirmed by spatial distribution analysis from classical MD simulations.

This study highlights that both mechanisms and rates of CO_2 absorption into aqueous amines can be strongly influenced by the entropic effects arising from steric constraints on surrounding H_2O molecules imposed by the geometry and chemical affinity of amines. The improved understanding at the molecular level may provide valuable guidance in the design of cost-effective amine-based solvents for CO_2 capture and utilization.

References

(1) IPCC. *Climate Change 2014: Synthesis Report. Contribution of Working Groups I, II and III to the Fifth Assessment Report of the Intergovernmental Panel on Climate Change*; Geneva, Switzerland, 2014.

(2) Rochelle, G. T. Amine Scrubbing for CO_2 Capture. *Science* **2009**, *325* (5948), 1652–1654.

(3) Kenarsari, S. D.; Yang, D.; Jiang, G.; Zhang, S.; Wang, J.; Russell, A. G.; Wei, Q.; Fan, M. Review of Recent Advances in Carbon Dioxide Separation and Capture. *RSC Adv.* **2013**, *3* (45), 22739–22773.

(4) Figueroa, J. D.; Fout, T.; Plasynski, S.; McIlvried, H.; Srivastava, R. D. Advances in CO_2 Capture technology—The U.S. Department of Energy's Carbon Sequestration Program. *Int. J. Greenh. Gas Control* **2008**, *2* (1), 9–20.

(5) Strazisar, B. R.; Anderson, R. R.; White, C. M. Degradation Pathways for Monoethanolamine in a CO_2

Capture Facility. *Energy and Fuels* **2003**, *17* (4), 1034–1039.

(6) Jakobsen, J. P.; Krane, J.; Svendsen, H. F. Liquid-Phase Composition Determination in CO_2 - H_2O - Alkanolamine Systems: An NMR Study. *Ind. Eng. Chem. Res.* **2005**, *44* (26), 9894–9903.

(7) Vaidya, P. D.; Kenig, E. Y. CO_2-Alkanolamine Reaction Kinetics: A Review of Recent Studies. *Chem. Eng. Technol.* **2007**, *30* (11), 1467–1474.

(8) Sartori, G.; Savage, D. W. Sterically Hindered Amines for CO_2 Removal from Gases. *Ind. Eng. Chem. Fundam.* **1983**, *22* (2), 239–249.

(9) Chakraborty, A. K.; Astarita, G.; Bischoff, K. B. CO_2 Absorption in Aqueous Solutions of Hindered Amines. *Chem. Eng. Sci.* **1986**, *41* (4), 997–1003.

(10) Lepaumier, H.; Picq, D.; Carrette, P.-L. New Amines for CO_2 Capture. I. Mechanisms of Amine Degradation in the Presence of CO_2. *Ind. Eng. Chem. Res.* **2009**, *48* (20), 9061–9067.

(11) Lepaumier, H.; Picq, D.; Carrette, P.-L. New Amines for CO_2 Capture. II. Oxidative Degradation Mechanisms. *Ind. Eng. Chem. Res.* **2009**, *48* (20), 9068–9075.

(12) Caplow, M. Kinetics of Carbamate Formation and Breakdown. *J. Am. Chem. Soc.* **1968**, *90* (24), 6795–6803.

(13) Danckwerts, P. V. The Reaction of CO_2 with Ethanolamines. *Chem. Eng. Sci.* **1979**, *34* (4), 443–446.

(14) Donaldson, T. L.; Nguyen, Y. N. Carbon Dioxide Reaction Kinetics and Transport in Aqueous Amine Membranes. *Ind. Eng. Chem. Fundam.* **1980**, *19* (3), 260–266.

(15) Versteeg, G. F.; Van Swaaij, W. P. M. On the Kinetics Between CO_2 and Alkanolamines Both in Aqueous and Non-Aqueous Solutions - II. Tertiary Amines. *Chem. Eng. Sci.* **1988**, *43* (3), 587–591.

(16) Stowe, H. M.; Hwang, G. S. Fundamental Understanding of CO_2 Capture and Regeneration in Aqueous Amines from First-Principles Studies: Recent Progress and Remaining Challenges. *Ind. Eng. Chem. Res.* **2017**, *56* (24), 6887–6899.

(17) Sharma, M. M. Kinetics of Reactions of Carbonyl Sulphide and Carbon Dioxide with Amines and Catalysis by Bronsted Bases of the Hydrolysis of COS. *Trans. Faraday Soc.* **1964**, *61* (6), 681–688.

(18) Kim, C. J.; Savage, D. W. Kinetics of Carbon Dioxide Reaction with Diethylaminoethanol in Aqueous Solution. *Chem. Eng. Sci.* **1987**, *42* (6), 1481–1487.

(19) Alper, E. Reaction Mechanism and Kinetics of Aqueous Solutions of 2-Amino-2-Methyl-1 -Propanol and Carbon Dioxide. *Ind. Eng. Chem. Res.* **1990**, *29* (8), 1725–1728.

(20) Stowe, H. M.; Vilčiauskas, L.; Paek, E.; Hwang, G. S. On the Origin of Preferred Bicarbonate Production from Carbon Dioxide (CO_2) Capture in Aqueous 2-Amino-2-Methyl-1-Propanol (AMP). *Phys. Chem. Chem. Phys.* **2015**, *17* (43), 29184–29192.

(21) Stowe, H. M.; Hwang, G. S. Molecular Insights into the Enhanced Rate of CO_2 Absorption to Produce Bicarbonate in Aqueous. *Phys. Chem. Chem. Phys.* **2017**, *19*, 32116–32124.

534958 CO_2-Based e-Fuels - a Comparative TEA of Methanol & Omes Based on a Novel Assessment Guideline

Abstract:

A major CO_2-utilising product group are fuels, also called electro fuels ('e-fuels'). E-fuels show three major benefits: Large-scale storage of fluctuating renewable energy, reduction of carbon intensity and reduction of pollutants for transport fuels (especially soot and NO_x).[1] E-fuels can become a major market with projected revenues of 10 - 250 billion USD in 2030, utilizing up to 2.1 Gt of CO_2.[2]

E-fuels R&D has progressed from lab-scale to pilot-scale and a range of industrial e-fuel projects dealing with methanol, ethanol and Fischer-Tropsch oil exist today.[2] But to replace diesel, methanol, ethanol and other alcohols are not suited due to too low cetane numbers.[1] Instead, Fischer-Tropsch diesel and oxymethylene ethers (OME) were identified as promising substitutes, especially dimethyl ethers (DME),[1] corresponding to "OME_0",[3] dimethoxymethane (DMM) corresponding to "OME_1",[4] and oligomeric OMEs with 3 to 5 repeating units (OME_{3-5}).[1]

While techno-economic (TEA) studies exist for CO_2 utilizing methanol, DME or FT-diesel, a detailed analysis of OME_1 and OME_{3-5} from various CO_2 sources has not been performed. In this work, we present a TEA of the production of four e-fuels: Methanol, DME/OME_0, DMM/OME_1 and OME_{3-5}. Various production routes will be analyzed, combining different CO_2 emitters (ammonia plants, steel plants, and cement plants), and different electricity routes (natural gas, grid and wind). Regional scenarios reflecting the US and EU conditions are further included. The process design software Aspen Plus is used to develop process design flow sheets in order to calculate mass and energy balances, efficiencies and economics. For the commercialization of e-fuels, standardized and fair comparisons are necessary. This is why this assessment is based on a novel, standardized TEA framework developed in the TEA Guideline project, integrating concepts from TEA and LCA to make studies more transparent and comparable.

The main objective of this work is first to exemplify the TEA framework and second to identify promising e-fuel production options for CO_2 utilization in the processing industry. The study will outline future e-fuel production scenarios and focus points for the next steps in R&D.

Acknowledgments

Support from EIT Climate-KIC and CO_2 Sciences for the project 'TEA Guidelines' is gratefully acknowledged.

References

[1] S. Schemme, R. C. Samsun, R. Peters, D. Stolten, *Fuel* **2017**, *205*, 198–221.

[2] CO2 Sciences, The Global CO2 Initiative, *Global Roadmap for Implementing CO2 Utilization*, **2016**.

[3] J. Artz, T. E. Müller, K. Thenert, J. Kleinekorte, R. Meys, A. Sternberg, A. Bardow, W. Leitner, *Chem. Rev.* **2017**, acs.chemrev.7b00435.

[4] S. Deutz, D. Bongartz, B. Heuser, A. Kätelhön, L. Schulze Langenhorst, A. Omari, M. Walters, J. Klankermayer, W. Leitner, A. Mitsos, et al., *Energy Environ. Sci.* **2018**, *11*, 331–343.

538955 A Macroscopic Model Accounting for the Composite Effects for an Ion Lithium Cell with a $LiFePO_4$ Cathode

Abstract:

The increasing energy demand and actuals pollution problems derived from fossil fuels require the generation of energy by renewable sources, and energy storage devices to solve the intermittency problems. Li-ion batteries have been positioned as one of the best storage devices recently in electric vehicles and have the possibility to implement as large energy storage [1]. A typical Ion-Lithium cell is formed for two electrodes (anode and cathode), one separator to avoid the contact between them, and an electrolytic solution containing dissociated salts for enable ion transfer between two electrodes [1-3].

The electrodes are porous structures composed by the mixture of the active material and additional compounds that improve the transport properties of charge and mass; however, it is paramount to account for numerous problems related to transport mechanisms of charge and mass that significantly affect its capacity and power [4]. The understanding of these mechanisms is complex due to various kinetic and transport phenomena interacting within the cell, and the difficulty of predicting local variables such as concentration, potential, among others. Mathematical modelling is a tool normally used to understand and decouple this interaction [5-10].

On other hands, the composition of the cathode slurry has, recently, been reported to influence battery performance and rate capability [10]. In order to understand the battery behaviour, the relation of Li^+ diffusion with cathode composition and operation conditions (C-rate and state of discharge), in this work, we propose a pseudo-heterogeneous model for the description of the electrochemical behaviour of a half cell configuration of Li°/1M $LiPF_6$ in 1:1:1 CE: DMC:EMC/$LiFePO_4$:PVDF-Csp. This model accounts for the main transport phenomena involved in the cell (diffusion and migration) and the intercalation kinetics of Li^+ ion in the active material from electrode. This model is used to analyse the influence of operation conditions in the transport and kinetic mechanisms during the cell operation.

References

1. Scrosati, B. and J. Garche, Lithium *batteries: Status, prospects and future.* Journal of Power Sources, 2010. **195**(9): p. 2419-2430.
2. Goodenough, J.B. and K.-S. Park, *The Li-Ion Rechargeable Battery: A Perspective.* Journal of the American Chemical Society, 2013. **135**(4): p. 1167-1176.
3. Ibrahim, H., A. Ilinca, and J. Perron, *Energy storage systemsâ€"Characteristics and comparisons.* Renewable and Sustainable Energy Reviews, 2008. **12**(5): p. 1221-1250.
4. GuzmÃ¡n g., et al., Improved performance of LiFePO4 cathode for Li-ion batteries through percolation studies. Electrochimica Acta, 2017. **247** p. 451-459
5. Park, M., et al., *A review of conduction phenomena in Li-ion batteries.* Journal of Power Sources, 2010. **195**(24): p. 7904-7929.
6. Farkhondeh, M. and C. Delacourt, *Mathematical Modeling of Commercial LiFePO 4 Electrodes Based on Variable Solid-State Diffusivity*. Vol. 159. 2012. 177-192.
7. Farkhondeh, M., et al., *Mesoscopic Modeling of a LiFePO 4 Electrode: Experimental Validation under Continuous and Intermittent Operating Conditions*. Vol. 164. 2017. E3040-E3053.
8. Farkhondeh, M., et al., *Full-Range Simulation of a Commercial LiFePO4 Electrode Accounting for Bulk and Surface Effects: A Comparative Analysis.* Journal of The Electrochemical Society, 2014. **161**(3): p. A201-A212.
9. Safari, M. and C. Delacourt, *Mathematical Modeling of Lithium Iron Phosphate Electrode: Galvanostatic Charge/Discharge and Path Dependence.* Journal of The Electrochemical Society, 2011. **158**(2): p. A63-A73.
10. Mastali Majdabadi, M., et al., *Simplified electrochemical multi-particle model for LiFePO4 cathodes in lithium-ion batteries.* Journal of Power Sources, 2015. **275**: p. 633-643.

Acknowledgements

I.O. Santos Mendoza thanks to CONACYT for the scholarship granted to pursue her doctoral studies and Universidad Autónoma Metropolitana for the support.

540743 Research of the Vertical Falling Film Behavior in the Scrubbing-Cooling Tube

Abstract:

Research on the vertical falling film behavior in the scrubbing-cooling tube

Yifei Wang*, Xin Peng, Liucheng Yan , Guangsuo Yu, Fuchen Wang

Key Laboratory of Coal Gasification and Energy Chemical Engineering of the Ministry of Education, East China University of Science and Technology, Shanghai 200237, PR China

Being a part of the opposed muti-burner (OMB) gasifier, the scrubbing-cooling chamber is used to cool, humidify and scrub the high temperature raw syngas. The chamber mainly consists of a scrubbing-cooling ring, a scrubbing-cooling tube, a liquid pool and a bubble bed. The cooling water flows through the ring and forms vertical liquid film inside the tube. The liquid film protects the wall of the tube from the high temperature syngas and slag. Most of the falling film studies are carried out in micro-scale tube with small liquid Reynolds number. The research here focuses on the study of both of the velocity and thickness distribution of falling liquid film under relatively large Reynolds number which is closed to actual industrial condition.

An Ultrasound Doppler Velocimetry was employed to measure the velocity and thickness distribution of falling liquid film and the liquid Reynolds number ranged from 1.0×10^4 to 3.1×10^4. A 3D numerical model of the scrubbing-cooling tube was established and the simulation results were in good agreement with the experimental data as shown in Fig. 1. The average liquid film thickness became largest at the 0 circumferential position while a part of the liquid was separated from the local liquid film and dripped in the gas core when Reynolds number was larger than 1.5×10^4 as shown in Fig. 2-3. In the lower section of the tube, the circumferential liquid film thickness distributed more evenly. With the increase of Reynolds number, both of the thickness and velocity of global liquid film increased and the thickness distribution became unevener. Under the effect of gravity, the liquid film velocity increased with the increasing flow distance at small Reynolds numbers while decreasing at large Reynolds numbers due to the resistance.

Keywords: falling film, film velocity and thickness, numerical simulation, large Reynolds numbers, Ultrasound Doppler Velocimetry

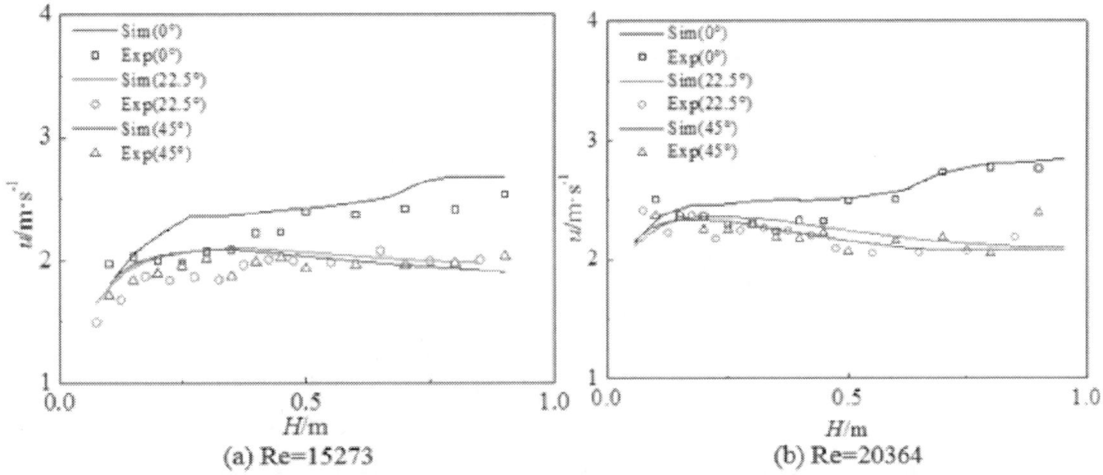

Fig. 1 Comparison of liquid film velocity distribution between simulated and experimental data

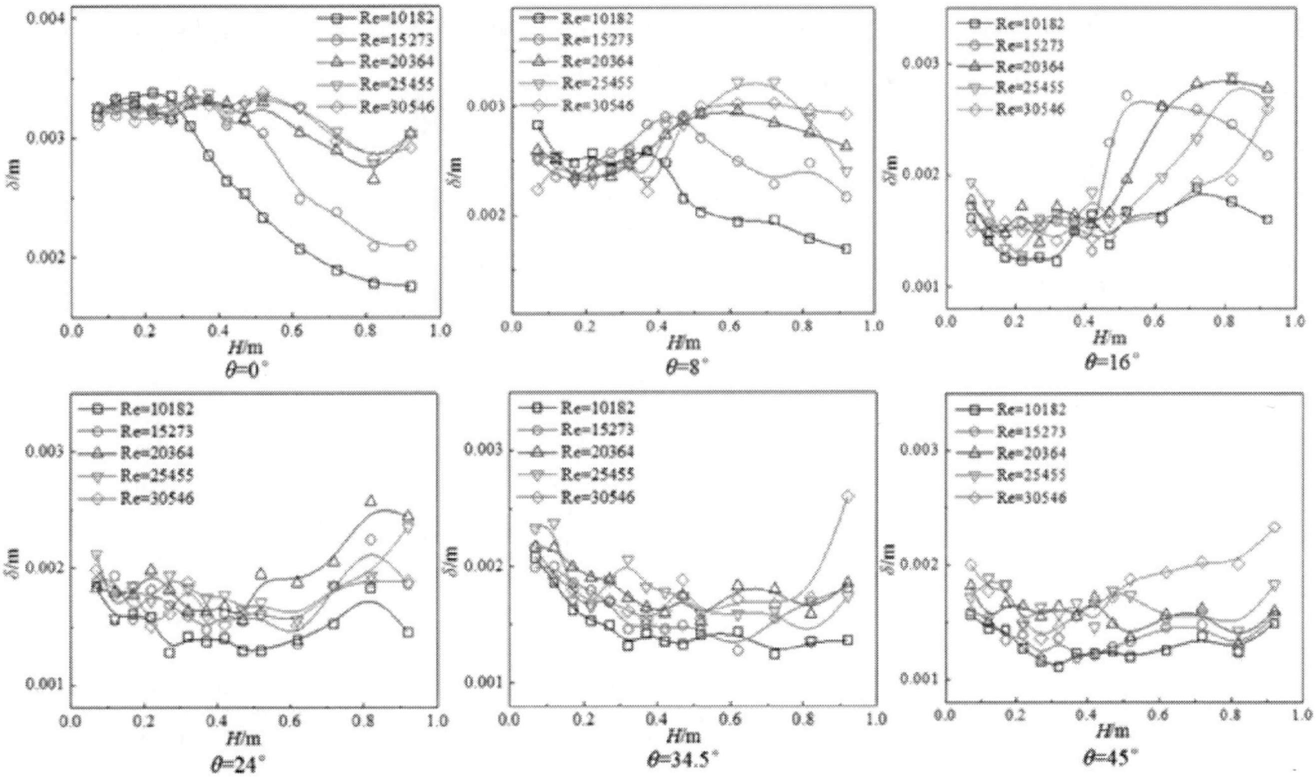

Fig. 2 Liquid film thickness distribution at different Reynolds numbers

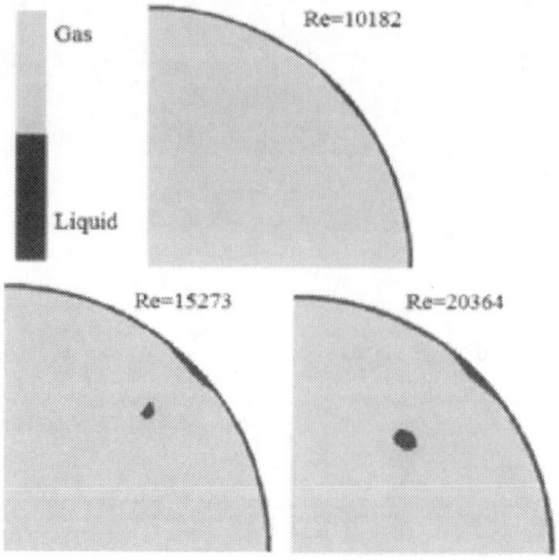

Fig. 3 Phase contour graph of the outlet of the tube

* Corresponding Author: Tel: +86 21 6425 2522. Fax: +86 21 6425 1312. E-mail: wangyf@ecust.edu.cn

527141 Effect of Shear Rate and Drying Speed in Lithium Ion Battery Slurry Processing

Abstract:

Processing conditions of battery slurries into electrodes are known to affect final battery performance. However, there is a lack of fundamental understanding of how to control processing conditions to achieve better batteries. Previous work has focused on the formulation step of the battery manufacturing process. This study concentrates on two of the slurry processing steps, namely coating and drying, and their effect on film quality and electrode performance. Rheological measurements were used to determine the starting slurry microstructure as well as determining the effect of flow on microstructure evolution. Well-characterized slurries were subjected to a series of shear rates and drying temperatures to examine the effect of flow and drying rate on final electrode performance. The data suggests that there is a complex relationship between surface roughness and shear rate/temperature, but ultimately, performance based on shear rate is dependent on drying time. Wet slurries are well mixed to form homogeneous suspensions. The slurries are spread via an automatic bird applicator coater to the desired thickness. The film is quickly exposed to different baking temperatures to remove solvent. We observe that higher temperatures produce better performing batteries on average. We argue that quicker drying times prevent sedimentation and phase separation of the slurry. The shear rate has a unique influence on the final electrode performance, and the optimum shear rate depends on the temperature if drying. We use discharge capacity data, scanning electron microscopy and energy dispersive spectroscopy studies to characterize the performance and final electrode particle and chemical species dispersion, respectively. While more data is needed to make concrete conclusions, the preliminary data presented here shows for the first-time quantitative relationships between processing conditions and battery performance.

538518 Improving Electrode Performances for Air-Cathode Microbial Fuel Cells

Abstract:

Microbial fuel cells (MFCs) is a bio-electrochemical system that utilizes the microorganisms and organic wastes to generate electrical energy. The microorganisms are exoelectrogenic bacteria that serve as both catalyst and electron transporter. In this study, the exoelectrogenic bacteria were attached and cultivated on the graphite electrode. The graphite electrodes were further modified to increase surface area and to improve hydrophilicity. Linear sweep voltammetry and cyclic voltammetry analyses indicated that the modified electrode could improve electrochemical properties. The modified electrode in a half H-type bottle microbial fuel cell (H-MFC) with a Pt-coated carbon paper as both counter and reference electrode had a higher power density, which was more 100% than the unmodified electrode. Besides, by adjusting the electrode distance and increasing the effective area of cathode, a higher power density could be reached. This work demonstrated the preparation of a hydrophilic graphite electrode to improve performance microbial fuel cells.

SOLAR HYBRID POWER SYSTEMS FOR IMPROVED ELECTRICAL DELIVERY IN LOW HUMIDITY, HIGH TEMPERATURE CONDITIONS

Mounir Bouzguenda[1], Aly A. Aboulnaga[1], Gafar Elamin[2], Dillon M. Hayman[3], Kenneth L. Roberts[3]

[1] - Department of Electrical Engineering
King Faisal University
Hofuf, Al Ahsa, Kingdom of Saudi Arabia

[2] - Department of Mechanical Engineering
King Faisal University
Hofuf, Al Ahsa, Kingdom of Saudi Arabia

[3] - Department of Chemical Engineering
University of South Carolina
Columbia, SC, U.S.A.

Summary

The implementation of 50% of electricity from non-hydrocarbon resources in the Kingdom of Saudi Arabia by 2032 with 54 GW from renewable energy sources, including 41 GW from photovoltaics (PV) and concentrated solar power (CSP), 9 GW wind turbine power, 3 GW waste-to-energy conversion, and 1 GW geothermal power, are national targets[1]. Solar energy collection is a leading source of renewable power technology concurrently with wind turbines, geothermal, and low temperature PEM fuel cells that are currently under investigation in Saudi Arabia. Wind turbines can also provide a steady source of sunlight-related power, but this technology grapples with issues ranging from with community building regulation, incompatibility with aircraft flight zones and noise pollution issues. Geothermal energy is another viable alternative renewable energy source that unfortunately deals with issues of environmental regulation and unknown factors dealing with effects on living communities with hydrogen sulfide, sulfur dioxide and toxic metals, akin to fracking. As mentioned previously, low temperature PEM fuel cells can effectively provide high efficiency power but for relatively limited consumer applications and within limited ambient operating conditions. Long term operation of PEM fuel cells in high temperature, arid conditions has been reported to operational failure of these devices due to the drying of electrolyte[2]. With the large amounts of available sunlight and relatively optimal weather conditions in Saudi Arabia, the continued development of solar power-based systems presents the most promising pathway for large-scale renewable power in KSA.

The inconsistent availability of solar energy during peak load times after sundown, cloud-obstructed daytime hours and in highly dusty conditions has also led to the recent investigation of hybrid solar-PV energy systems using secondary renewable and/or nonrenewable energy sources[3-5]. The advantages of hybrid power systems include the potential availability of two or more power sources to compensate for time periods of a single energy source shortages, lower maintenance costs and improved load matching[3-5]. Disadvantages include decreases in the overall efficiency of energy conversion using multiple power sources, the higher cost of additional renewable energy sources and the environmental effects of using secondary nonrenewable power sources with solar energy technology.

This project seeks to assess and evaluate the viability of developing hybrid solar PV-fuel cell systems for supplying electricity customers in Saudi Arabia and similar climates during times of minimal sunlight and especially during peak load hours (i.e., afternoon / evening hours in the Summer months) to reduce the risk of blackouts and power grid failures. A new renewable hybrid engineering model has been developed for a solar PV-thermoelectric generator hybrid system to determine the optimal parameters for operation and for design of a prototype.

Introduction

The implementation of 50% of electricity from non-hydrocarbon resources in the Kingdom of Saudi Arabia by 2032 with 54 GW from renewable energy sources, including 41 GW from photovoltaics (PV) and concentrated solar power (CSP), 9 GW wind turbine power, 3 GW waste-to-energy conversion, and 1 GW geothermal power, are national targets[1]. Solar energy collection is a leading source of renewable power technology concurrently with wind turbines, geothermal, and low temperature PEM fuel cells that are currently under investigation in Saudi Arabia. Wind turbines can also provide a steady source of sunlight-related power, but this technology grapples with issues ranging from with community building regulation, incompatibility with aircraft flight zones and noise pollution issues. Geothermal energy is another viable alternative renewable energy source that unfortunately deals with issues of environmental regulation and unknown factors dealing with effects on living communities with hydrogen sulfide, sulfur dioxide and toxic metals, akin to fracking. As mentioned previously, low temperature PEM fuel cells can effectively provide high efficiency power but for relatively limited consumer applications and within limited ambient operating conditions. Long term operation of PEM fuel cells in high temperature, arid conditions has been reported to operational failure of these devices due to the drying of electrolyte[2]. With the large amounts of available sunlight and relatively optimal weather conditions in Saudi Arabia, the continued development of solar power-based systems presents the most promising pathway for large-scale renewable power in KSA.

The implementation of solar energy-based systems for the 24-hour production of electricity also requires the consideration of effective hybrid systems in tandem with photovoltaic technology. For small housing needs, the use of PEM fuel cells with battery storage is of great interest. However, for larger PV applications to provide electricity to cities and towns, much larger kW capacity (50+ kW) energy storage / delivery systems must be developed to supply electricity consumers during times of minimal sunlight. Currently the development of hybrid PV-TEG and solar-fuel cell systems is globally of great interest for the generation of electricity needs, especially in remote areas[6]. Fuel cells have extremely higher efficiency of operation than conventional power generation technologies (30-65% for electricity production) and when used in tandem with hydrogen storage systems can in theoretically provide continuous power as long as a reliable hydrogen supply is available. As a renewable energy source, fuel cells can operate using numerous sources of hydrogen including fossil fuels, liquid fuels, carbonaceous waste products and others, with only electricity and water vapor being the primary products of operation[7].

There are several technology issues with the effective implantation of solar energy collection technology in Saudi Arabia including the need for solar panel self-cleaning technologies, understanding the compatibility of solar cell stations with existing power grid technologies, optimizing power conversion & electronics for PV systems, and the development of reliable energy storage technologies to ensure a constant power supply during periods of minimal sunlight. For solar power systems, the storage of additional collection solar energy has been a primary focus with the proposal of high capacity battery storage systems, thermal energy systems, compressed air turbine systems and fuel cell-hydrogen systems as leading technologies for power storage supplements to solar power systems.

The primary research in hybrid solar-fuel cell technology involves the utilization of a PEM fuel cell (PEMFC). However, ongoing research has shown that PEM fuel cells have operational issues in high temperature, arid environments that lead to the severe drying of the electrolyte materials and failure of this technology. The PEMFC also provides relatively low electricity capacity (<5-10 kW) and is typically designed for family homes or small-scale business sites. In the high temperature, arid environments of the Arabian Peninsula, the use of high temperature fuel cell technologies such the Solid Oxide fuel cell (SOFC). The SOFC can operate effectively in high temperature, arid conditions by providing power capabilities in excess of 200 kW with an overall efficiency of 40% electrical output and 65% electrical-cogeneration heating output[7]. As was mentioned above, solid oxide fuel cells can operate using various feedstock including pure hydrogen, natural gas, liquid hydrocarbon-based fuels and even carbonaceous waste materials.

The development of solar PV-PEMFC-hydrogen system has been recently reported by Santoso and coworkers[8]. This work reported the monitoring of a PEM Electrolyzer-Solar Cell-PEM Fuel Cell-Hydrogen system that utilized electricity from the PEMFC to electrolyze water into hydrogen and oxygen. The fuel cell system consisted of 13 cells that had capability to generate 20 W of power with rated performance of 7.8 V at 2.6 A. Excess hydrogen was stored using a metal hydride storage unit. Maximum power output for this system in series and parallel configuration was between 15 to 45 W. Similar systems have been reported by several other groups[6,9-11]. Another opportunity with using fuel cell technology is the availability of hydrogen. As the largest supplier of oil and natural gas in the world, the Kingdom of Saudi Arabia has more than sufficient capacity to create a hydrogen infrastructure on par with those operating in the State of California in the United States and the Shell Oil-sponsored infrastructure technology in Holland[12-13]. Even without this hydrogen infrastructure, a viable alternative for storing and supplying hydrogen to fuel cells is the metal hydride system.

This work has investigated the theoretical use and efficiency of a photovoltaic-thermoelectric hybrid system. Average temperature data is used to determine ambient temperature throughout the year from 1981-2010. In addition, change in the sun's irradiation is modeled generally for a standard calendar year. This paper is intended to model a Poly-Si solar cell, Bi_2Te_3 thermoelectric hybrid system. A 1D, steady-state heat transfer model and thermodynamic analysis is applied to find theoretical efficiencies of both the solar PV component and thermoelectric component of the system. This model assumes constant properties throughout each material. Power and efficiency have been calculated individually for both materials. Total power output of the system throughout the year was also determined and compared to a stand-alone PV system. When a photon in the proper band of light energy hits the photovoltaic material, an electron in the valence shell of the atoms in the material are excited to a higher energy state. When the electrons move, they leave a "hole" in the valence shell. The absence of electrons in the valence shells causes electrons from the negative layer to move to fill the space creating an electric current as they travel. Today around 90% of solar panels are Si-based which harvest energy in the visible light spectrum and somewhat into the infrared spectrum[14].

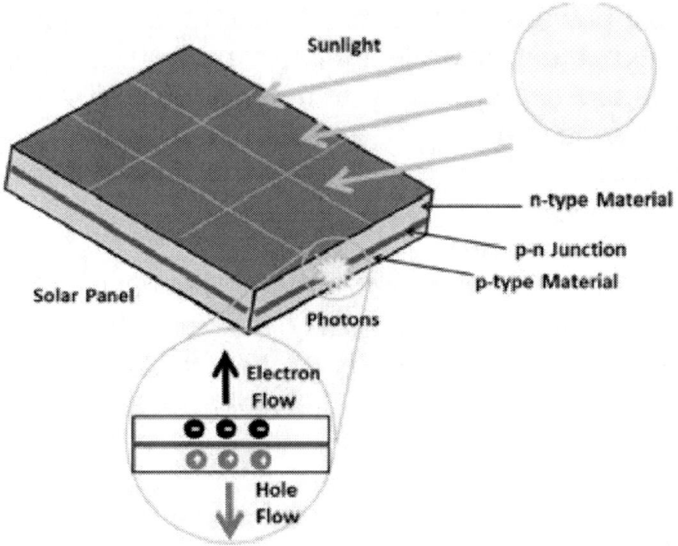

Figure 1: *Diagram of the photoelectric effect* [8]

Thermoelectric materials on the other hand develop an electric potential when positively and negatively separate upon the material being heated. This property is known as the Seebeck effect and creates a unique opportunity to harvest energy in the thermal spectrum. An effective thermoelectric material has a high electrical conductivity but a low thermal conductivity so that a gradient between the hot and cold side can develop. A dimensionless parameter "Z" is used to evaluate thermoelectric materials. It was reported via a combination of the materials electrical and thermal conductivities as well as the Seebeck coefficient of a given material[15]. A common value for the "Z" for bismuth telluride materials was modeled in this work as 1 [16].

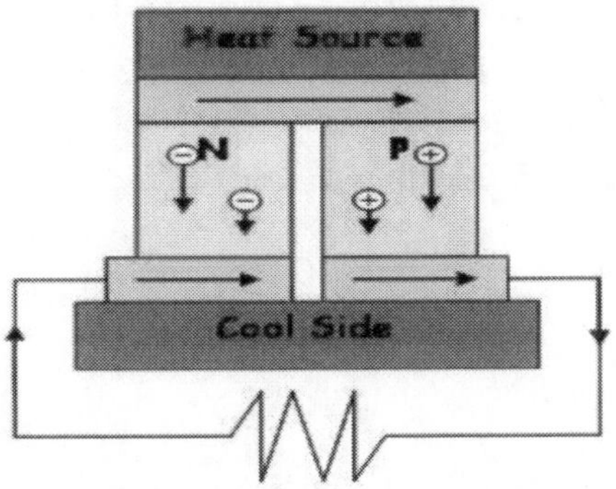

Figure 2: *Diagram of a thermoelectric generator*[17]

Results and Discussion

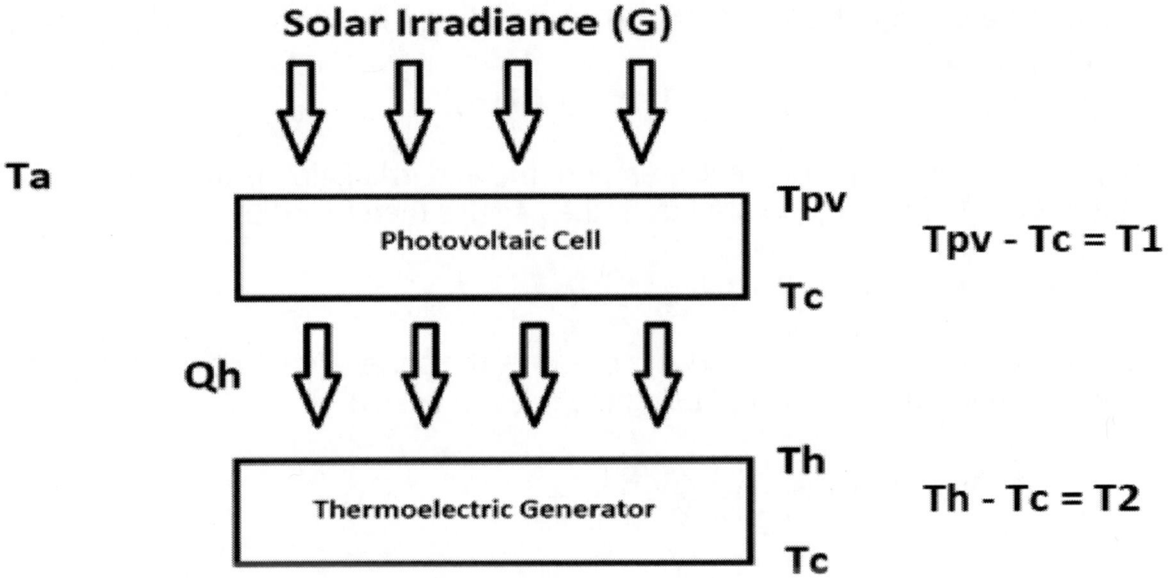

Figure 3: *Diagram of a PV –TEG Hybrid System*[15]

Temperature at the surface of the PV panel is assumed to be the product of ambient temperature (T_a) and solar irradiation (G).

$$T_{PV} = T_a + G \tag{1}$$

Solar irradiance (G) is calculated using a general modeling equation for irradiation throughout a year.

$$G = E_o * \{1+.034*cos[(360/360.25)*(Z\text{-}1)]\} \tag{2}$$

where E_o is the solar irradiation constant taken to be 1367 W/m^2 and Z represents the calendar day of interest from 1 to 365 days. Once the temperature at the surface of the panel is verified, the temperature at the bottom can be found by Fourier's law. After this value is combined with ambient temperature data, the properties of the thermoelectric material can be elucidated via Fourier's Law:

$$k_1 * A * \frac{dT_1}{dX_1} = k_2 * A * \frac{dT_2}{dX_2} \qquad (3)$$

where k_1 and k_2 are the respective thermal conductivities, X_1 and X_2 are the thicknesses and T_1 and T_2 are the changes in temperature of the materials. By solving these equations, the change in temperature of the thermoelectric material can be found. The temperature at the top of the thermoelectric material is considered to be the product of T_{PV} and the change in temperature. The heat flowing through the thermoelectric (Q_h) can also be found by another application of Fourier's Law.

The efficiencies of both the PV cell and TE material can be elucidated. The efficiency of the PV layer is found via:

$$\eta_{PV(Ta, G)} = \eta_{PV(G, TREF)} * [1 - B*(T_{PV} - T_{REF})] \qquad (4)$$

The efficiency of the panel at the reference temperature and irradiation is taken to be 12.4% and B equal to 0.004 [15]. Power generated from the panel is then found by:

$$P_{PV} = \eta_{PV\,(Ta, G)} * G * A \qquad (5)$$

where G is calculated from Equation 2 and A represents the panel area. The efficiency of the thermoelectric material can be estimated using the relation below:

$$\eta_{TE} = \frac{T_h - T_c}{T_h} * \frac{\sqrt{1 + Z*T} - 1}{\sqrt{1 + Z*T} + T_c/T_h} \qquad (6)$$

where $T = 0.5*(T_h + T_c)$. Additionally, T_c is assumed equal to the ambient temperature, T_h is the

temperature at the top of the thermoelectric material and Z is assumed to be 1.

After finding the efficiency of the thermoelectric material, the power generated can be estimated using:

$$P_{TE} = Q_h * \eta_{TE} \qquad (7)$$

Total power of the hybrid system is equivalent to the following relationship,

$$P_{TOT} = P_{TE} + P_{PV} \qquad (8)$$

and total efficiency by,

$$\eta_{Total} = P_{TOT} / (G * A) \quad (9)$$

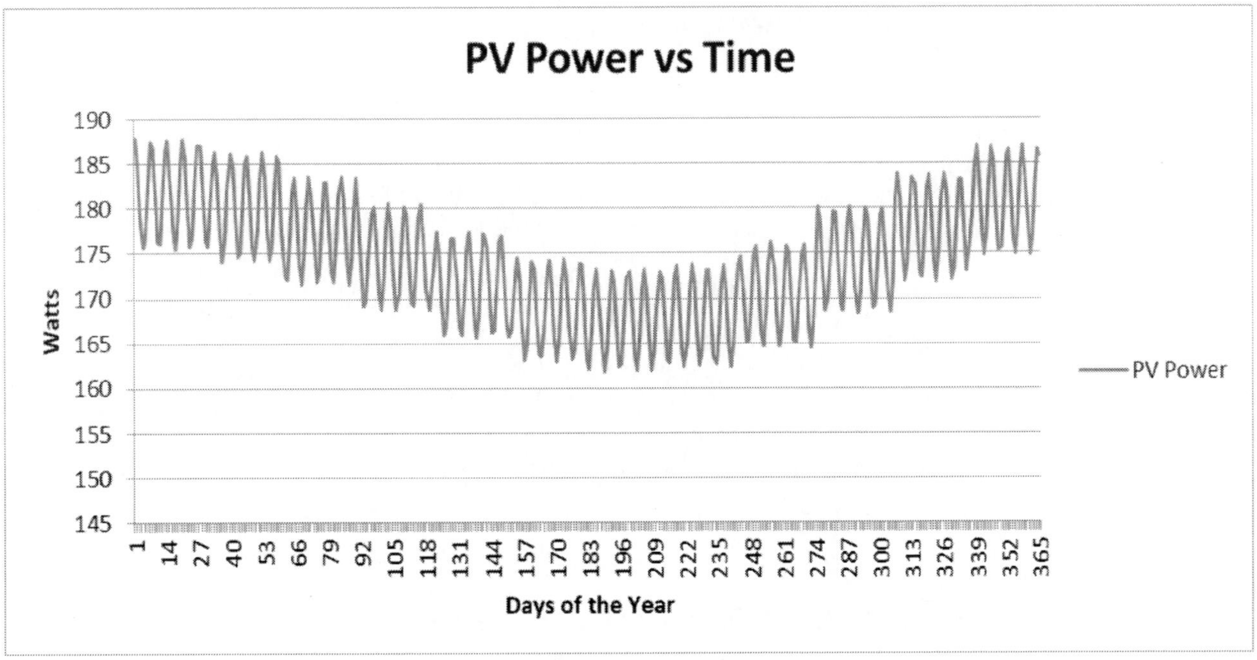

***Figure 4**: PV Power over Time*

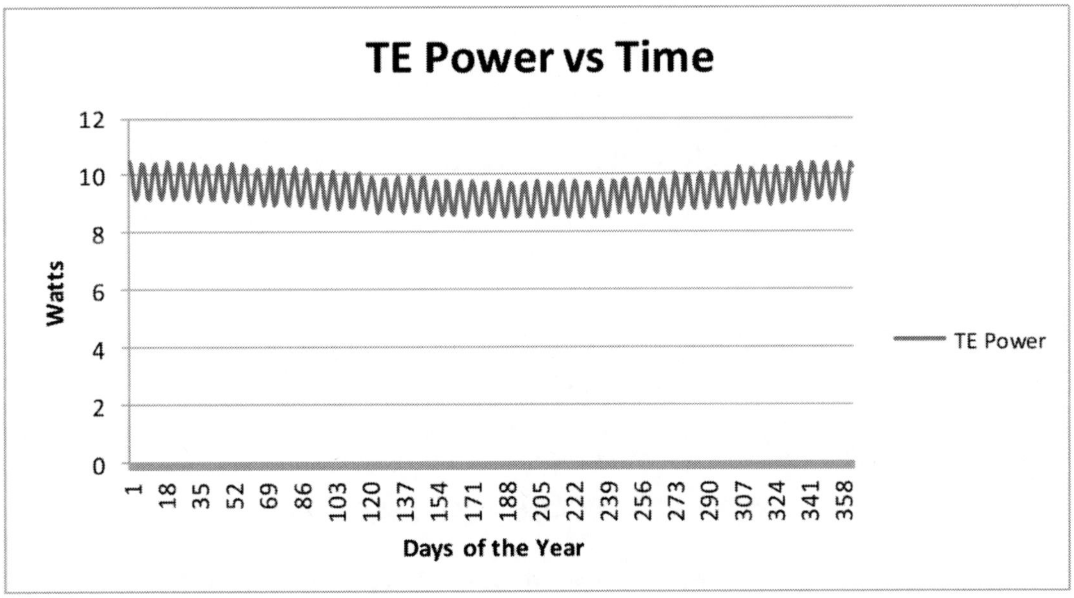

***Figure 5**: TE Power over Time*

As indicated in Figures 4 and 5, power for this model system would tend to decrease in the hotter months of the year. Some of form of heat management may be important to improve power output in the hotter months of the year. Thermoelectric power output tends to be more stable throughout the year which may help to buffer the larger dip in the power output for the PV cell.

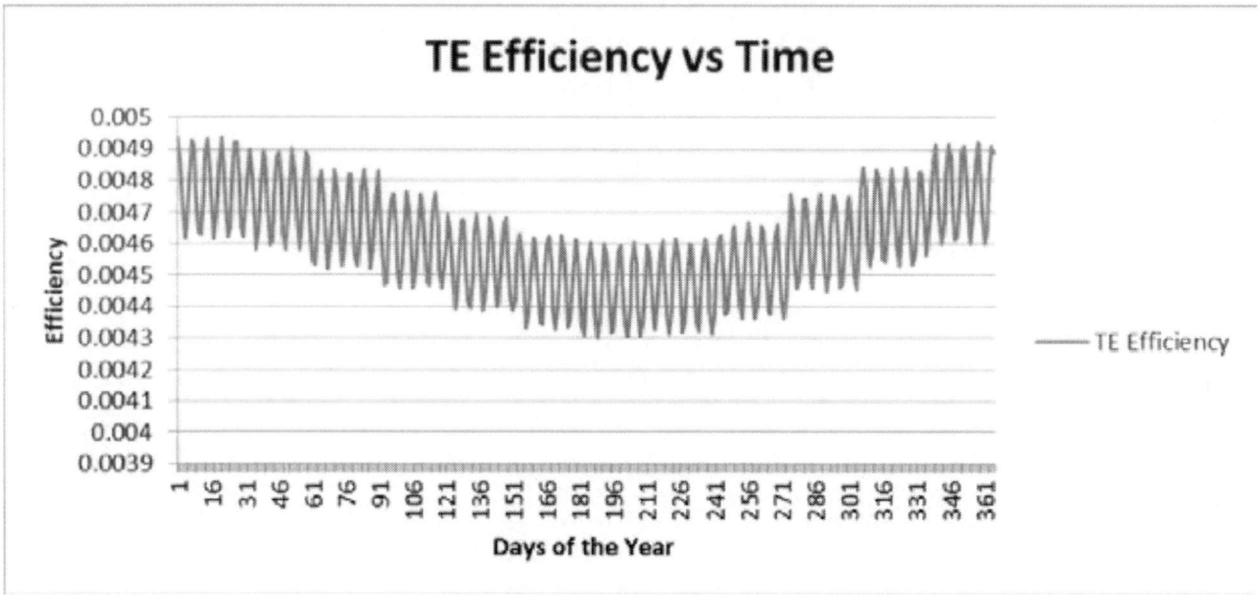

Figure 6: *TE Efficiency over Time*

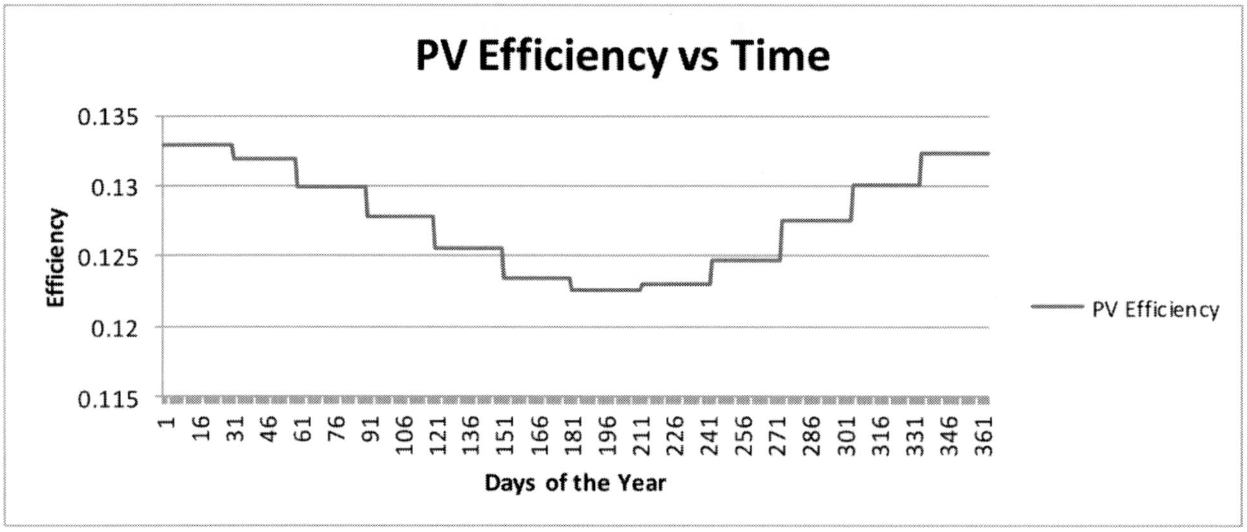

Figure 7: *PV Efficiency over Time*

The dip in power shown in Figures 4 and 5 is elucidated by the drop-in system inefficiency previously shown in Figures 6 and 7. Ambient temperature data was taken as a constant average over each month corresponding to the sharp drops and rises in Figure 4. However, ambient temperature seems to have less effect on the thermoelectric material.

Fluctuating solar irradiation seems to only have a large impact on the efficiency of the thermoelectric material.

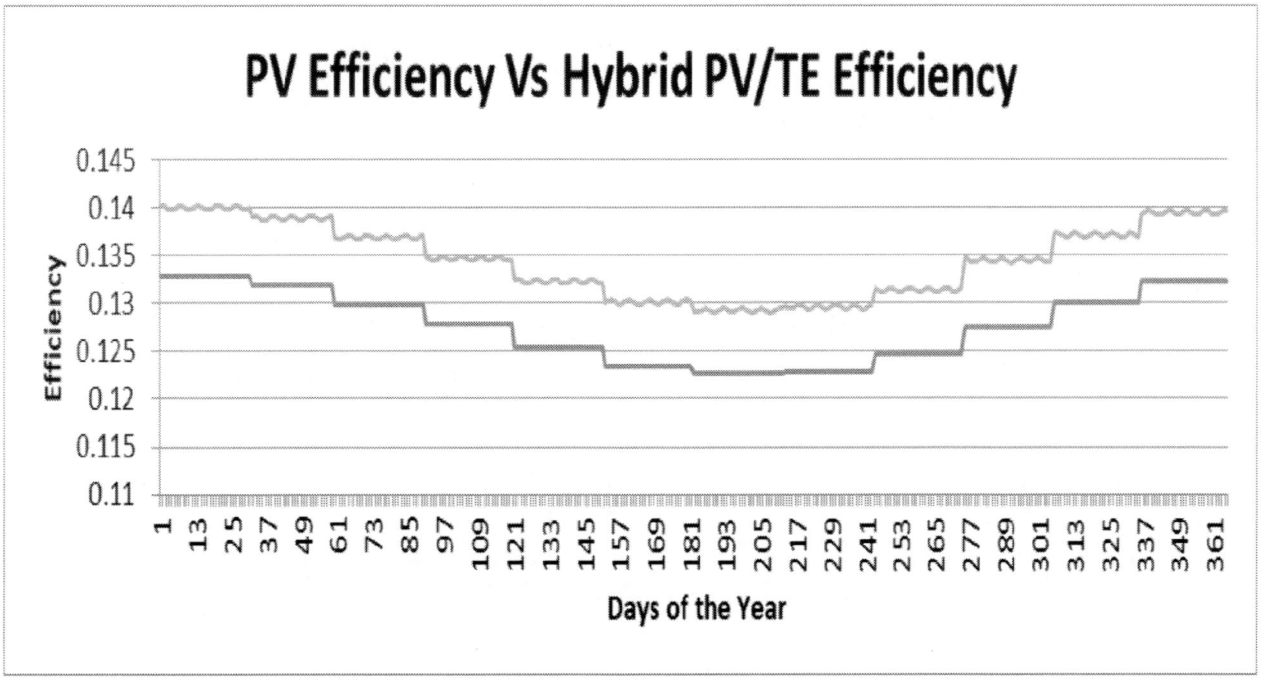

***Figure 8**: PV Efficiency (BLUE Curve) Versus Hybrid PV/TE Efficiency (RED Curve)*

Figure 8 shows the hybrid system efficiency in red and the stand-alone efficiency in blue. Efficiency of the hybrid system stays elevated above the stand-alone system throughout the year.

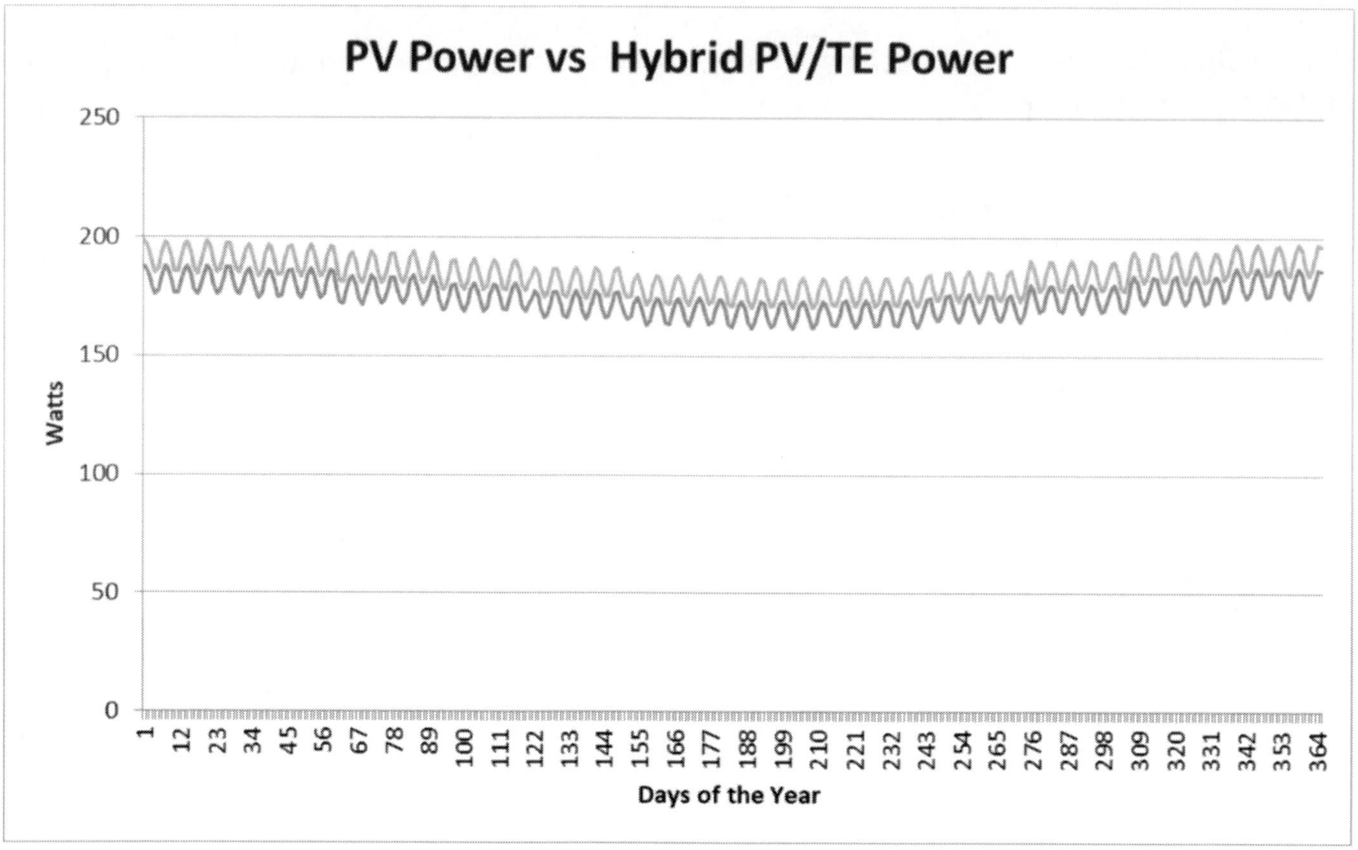

***Figure 9**: PV Power (BLUE Curve) vs Hybrid PV/TE Power (RED Curve)*

Figure 9 shows the increase in power output by adding the thermoelectric component to the system. Once again, the hybrid system is in red while the PV stand-alone system is in blue. The consistent power output of the thermoelectric generator as shown in Figure 5 indicates a power output that smoothly transitions in a gradual fashion over the year. The sine wave appearance of Figures 4, 5, and 8 are created by the fluctuating solar irradiation throughout the year modeled by Equation 2.

Conclusions

Photovoltaic technology shows promise for the future of renewable energy, but it will need additional modifications to be economical in the current market. Thermoelectric materials may offer a consistent source of power to help improve the overall efficiency of systems in the hotter months of the year. Future research into these systems may investigate cooling techniques to both pull heat from the solar panel and increase the heat difference across the thermoelectric material. In addition, future research in high performance thermoelectric nanomaterials can yield ZT values as high as 1.47 [18] as compared with a value of 1 used in this model. Additionally, the incorporation of thermal energy storage (TES) systems using such varied media as phase change materials (PCM), molten salts, and silica-based particulates &

monolithic structures would allow for a re-use of unused thermal energy for improved TE subunit performance[19]. In summary, the introduction of materials to utilize more of the available energy from the sun may lead to new hybrid systems such as the PV-TE system modeled in this paper which will bring PV technology to commercial viability.

References

1. El-Katiri, L. (2014) A Roadmap for Renewable Energy in the Middle East and North Africa, Oxford Institute for Energy Studies, Oxford, UK.

2. Yousfi-Steiner, N., Mocotéguy, Ph., Candusso, D., Hissel, D. (2009). A review on polymer electrolyte membrane fuel cell catalyst degradation and starvation issues: Causes, consequences and diagnostic for mitigation, *Journal of Power Sources*, v.194, pp. 130–145.

3. Akikur, R. K., Saidur, R., Ping, H.W., Ullah, K.R. (2015). Performance analysis of a co-generation system using solar energy and SOFC technology, *Energy Conversion and Management*, Vol. 79, pp. 415-430.

4. Bouzguenda, M. (2012). A comparative study of hybrid diesel solar PV–wind power systems in rural areas in the Sultanate of Oman, *International Journal of Sustainable Energy*, Vol. 31, Issue 2, pp. 95-106.

5. Sadeghi, S., Ameri, M. (2013). Multiobjective Optimization of PV-Bat-SOFC Hybrid System: Effect of Different Fuels Used in Solid Oxide Fuel Cell, *Journal of Energy Engineering*, v. 140, pp. 1-12.

6. Ural, Z., Gencoglu, M.T. (2014). Design and simulation of a solar-hydrogen system for different situations, *International Journal of Hydrogen Energy*, v. 39, pp. 8833-8840.

7. E.G. & G. Technical Services, Inc. (2004) *Fuel Cell Handbook, 7th Edition*, U.S. Department of Energy, Morganton, W.Va.

8. Santoso, D., Setiaji, F.D., Susilo, D. (2011). Demonstration of Renewable Electrical Energy Generation based on Solar-Hydrogen Fuel Cell Technology, *International Conference on Instrumentation, Communication, Information Technology and Biomedical Engineering*, 8-9 November 2011, Bandung, Indonesia.

9. Tarique, M.A., Dincer, I., Zamfirescu, C. (2014). Energy and Exergy Analyses of Solar-Driven ORC Integrated with Fuel Cells and Electrolyser for Hydrogen and Power Production, Dincer et al. (eds.), *Progress in Exergy, Energy, and the Environment*, pp. 81-89, Springer International Publishing, Switzerland.

10. Hisatomi, T., Kubota, J., Domen, K. (2014). Recent advances in semiconductors for photocatalytic and photoelectrochemical water splitting, *Chemical Society Reviews,* v. 43, pp. 7520-7535.

11. Ghribi, D., Khelifa, A., Diaf, S., Belhamel, M. (2013). Study of hydrogen production system by using PV solar energy and PEM electrolyser in Algeria, *International Journal of Hydrogen Energy*, v. 38, pp. 8480-8490.

12. Vagus, S. (26 February 2015). "California Invests $20 Million in Hydrogen Fuel Infrastructure", *Hydrogen Fuel News*, http://www.hydrogenfuelnews.com/california-invests-20-million-in-hydrogen-fuel-infrastructure/8521317/

13. Pyper, J. (7 November 2014). "Is Shell Boosting its Presence in Cleantech?", *greentechmedia*, http://www.greentechmedia.com/articles/read/Is-Shell-Boosting-Its-Presence-In-Cleantech

14. Greacen, C. (1991) How Photovoltaic Cells Work. *Home Power #23*, June-July, pg. 37.

15. Kossyvakis, D.N., Voutsinas, G.D., E.V. Hristoforou, E.V. (2016) Experimental analysis and performance evaluation of a tandem photovoltaic–thermoelectric hybrid system. *Energy Conversion and Management*, vol. 117, pp. 490–500.

16. Rezania, A., Sera, D., Rosendahl, L.A. (2016) Coupled thermal model of photovoltaic-thermoelectric hybrid panel for sample cities in Europe. *Renewable Energy*, vol. 99, pp. 127-135 (2016).

17. Brazier, K. (2008, January 11). [Thermoelectric Generator Diagram]. Retrieved August 03, 2018, from https://en.wikipedia.org/wiki/Thermoelectric_effect

18. Li, J-F, Liu, W-S, Zhao, L-D, Zhou, M. (2010) High-performance nanostructured thermoelectric materials. *NPG Asia Materials*, vol. 2, pp. 152–158.

19. Alva, G. Liu, L., Huang, X., Fang. G. (2017) Thermal energy storage materials and systems for solar energy applications. *Renewable and Sustainable Energy Reviews*, vol. 68, pp. 693–706.

AICHE *Annual Meet 2018*
October 26– November 2, 2018

David L. Lawrence Convention Center, Pittsburgh,

Comparative Study of Graphene/Water Nanofluid in a Heat Exchanger System: Modelling and Simulation

Devendra Purbia[1], Akshya Khandelwal[2], Arvind Kumar Sharma[3*]

[1,2] *Students at Birla Institute of Technology and Science Pilani, Jhunjhunu Rajasthan (India),* [3] *Associate Professor at Birla Institute of Technology and Science Pilani, Jhunjhunu Rajasthan (India),*
[1] *h20170042@pilani.bits-pilani.ac.in,* [2] *akshay21aug@gmail.com,* [3] *arvinds@pilani.bits-pilani.ac.in*

Abstract

Nanofluid is one of the vital solution for the enrichment of the thermal performance of heat transfer systems but at the same time it also enhance pressure drop. The increase in pressure drop effects the overall performance of heat exchanger system. So developing a nanofluid with high thermal conductivity to counter the effect of equivalent pressure drop can provide great advantage to its application. Therefore, in this paper Graphene Oxide is used as a nanoparticle and is compared with existing nanofluids such as TiO_2 (Titanium Oxide) and Al_2O_3 (Aluminum Oxide). Thermal performance and pressure drop analysis, through modelling and simulation, has been done using MATLAB. The base fluid is pure water and concentrations of nanoparticles are 0.025 %, 0.05 %, 0.075 %, 0.1 % respectively. The two dimensional steady state momentum transfer and heat energy consumption by nanofluids acting as a cooling media are modeled and simulated to compare the outcomes with experimental data . The flow is considered to be turbulent in the Reynolds number range of 3000-40000. The parameters are taken in the temperature range of 20 to 70 degree Celsius, with L/D ratio of tube dimensions greater than 10. Various parameters like Nusselt number, Peclet number, Pressure drop, friction factor along with the convective heat transfer coefficient were calculated using simulation. It was found that the Graphene/water nanofluid can be used to enhance the thermal performance of heat exchangers without much compromising with the negative effect of high pressure drop. Simulated results found to be in good agreement with the experimental results. Therefore, the present combination of nanoparticle and fluid can be the solution for energy efficient heat transfer systems.
Keywords : Heat Transfer, Nanotechnology, Graphene/Water Nanofluid

Annotation

μ_{nf}	Viscosity of nanofluid (Pa.s)
μ_w	Viscosity of water (Pa.s)
ρ_{nf}	Density of nanofluid (Kg/m^3)
v	Velocity (m/s)
ρ_s	Density of nanoparticle (Kg/m^3)
ρ_w	Density of water (Kg/m^3)
C_{pnf}	Specific heat of nanofluid (KJ/Kg.K)
Φ	Volume fraction
C_{pw}	Specific heat of water (KJ/Kg.K)
K_{nf}	Thermal conductivity of nanofluid (W/m.K)
K_f	Thermal conductivity of base fluid (Water) (W/m.K)
Kp	Thermal conductivity of nanoparticle (W/m.K)
N_{unf}	Nusselt number for nanofluid
h_{nfd}	Heat transfer coefficient of nanofluid (W/m^2.K)
Re_{nf}	Reynolds Number for nanofluid
Pr_{nf}	Prandtl number for nanofluid
D	Diameter of tube (mm)
Pec	Peclet number for nanofluid
V_{av}	Average velocity (m/s)
D_p	Diameter of particle (Nm)
$\propto_{nf}$	Diffusivity of nanofluid (m^2/s)

f_{nf}	Friction Factor of nanofluid
L	Length of tube (m)
Δp_{nf}	Pressure Difference of nanofluid (Pa)
τ_{rz}	Shear stress in r-z (Pa)
μ	Viscosity (Pa.s)
V_r	Velocity in r (m/s)
V_z	Velocity in z (m/s)
$\tau_{\theta z}$	Shear stress in $\theta - z$ direction (Pa)
τ_{zz}	Shear stress in z-z direction (Pa)
V_θ	Velocity in θ direction (m/s)
P	Pressure (Pa)
T	Temperature (K)
C_p	Specific heat (KJ/Kg.K)
K	Thermal conductivity (W/m.K)
Nx	Number of grids in axial direction
Ny	Number of grids in radial direction
T_{hotout}	Outlet temperature of hot fluid (Shell Side) (K)
T_{hotin}	Inlet temperature of hot fluid (K)
$T_{coldout}$	Outlet temperature of cold fluid (K)
T_E	Thermal energy factor
P_E	Pressure energy factor

Introduction

From the day of evolution, the urge of human for better quality of life is continuously increasing. Search of energy began much early after the discovery of fire. This increase in sophistication led to the development of industries which resulted in exploitation of energy at much higher pace. This exploitation of energy also resulted to another major problem that is pollution. The global temperature is on rise and it has reached the maximum ever in 2015 [1]. Number of meetings have been specially organized in the world for the "carbon management" over last few decades. The situation forecasted in above meetings predicts that these non-conventional energy resources will continue to prevail for some more time on the global scale as electrification is now being concentrated in the developing countries. Looking for the non-conventional energy resources will include huge amount of investment and budget allocation. Some of the densely populated countries like India have large reserves of coal and thus it seems unrealistic to think that these reserves will not be exploited [2,3].

Huge amount of greenhouse gas emission is directly associated with the use of non-conventional energy resources of which major proportion is emitted by the industries. Despite many efforts laid by the industries to bring down the total energy consumption, its utilization is still high above the threshold values. To decrease this energy consumption, more research is required to develop efficient systems used in the industries. In this search for the energy efficient industrial equipment, research started on industrial heat exchangers which are used in nearly every type of industries. In the beginning, research was more focused on developing the design of heat exchangers but now the interest of the researchers have shifted to its intrinsic development. Heat exchangers first started with ethylene glycol, water and oil as the base fluids but now with the new technology advancements, nanofluid with nanosized particles is introduced in the heat exchanger systems. Introducing nanoparticles inside the fluid imparts certain specific properties to the fluid. These nanofluids have numerous applications today in both laboratory or industrial levels [4].

Efficient boiling, higher critical flux and effective thermal conductivity are the salient features of the nanofluids. But at the same time, addition of nanoparticles in fluid rises the pressure drop in the system. This pressure drop is directly associated with the pump power, hence it is undesirable. Kumar et al. [5] investigated the friction factor and heat transfer of shell in the heat exchanger with helically coiled tube using Al_2O_3/water nanofluid with volume concentration of 0.1% - 0.8%. They found that addition of nanoparticles in the base fluid augments the thermal conductivity along with the friction factor. Assael et al. [6] showed that various nanofluids with different oxides of nanoparticles have better thermo physical properties as compared to the base fluid (water). The density and thermal conductivity of the nanofluid increased by 25 % and 10 % respectively in comparison to water, however, heat capacity reduced up to 20 %. The study conducted by Duang- thongsuk and Wongwises [4] with

TiO_2/water nanofluid resulted in increase in heat transfer coefficient along with little rise in the pressure drop. The pressure drop and convective heat transfer coefficient under turbulent conditions inside the circular tube are reported in the works of Fotukian and Nasr Esfahany et al., The results show that the use of nanoparticles augmented the thermal conductivity as well as the pressure drop [7].

The above studies on the nanoparticles shows that although they have ability to improve the thermal property of base fluid they also increases the pressure drop. The adverse effect of pressure drop decreases the overall profitability and thus it is not economical at industrial level. Therefore, we need to find a nanoparticle with high thermal conductivity but one with less pressure drop associated. In this search of new nanoparticle scientific community discovered a new material properties of which attracted many scientists of the present century. The material named "Graphene" is soft and thin allotrope of single carbon atom arranged in the hexagonal lattice. It has remarkable properties in terms of thermal or electrical conductivity. Graphene nanoparticle is very useful and can be used in numerous heat transfer applications [8].

Considering various studies above on Graphene and nanofluids, it can be corroborated that the use of graphene nanofluid as a base fluid in the heat exchangers can be propitious. Therefore, in this work graphene nanofluids is used as the base fluid in the heat exchange system to study its effect on overall thermal performance and pressure drop. The effect of nanoparticle on results obtained by modelling and simulation are compared with the results obtained from existing nanoparticles such as TiO_2 (Titanium Oxide) and Al_2O_3 (Aluminum Oxide). The results obtained after simulation of 2-D steady state flow model are compared with the experimental results. Pressure drop, Nusselt number, friction factor, Peclet number and heat transfer coefficient were calculated. MS Excel and MATLAB were used because of their powerful iterative and plotting abilities.

Mathematical Formulation

The properties of nanofluids are relatively different from those of base fluid. Most of the properties like density, viscosity, heat transfer coefficient get disturbed when nanoparticle dispersed in it. The density, viscosity, heat transfer coefficient are calculated using the available correlations in the literature. To calculate the viscosity for nanofluid, Einstein equation is generally used. The equation is suggested by Drew and Passman [9]

$$\mu_{nf} = (1 + 2.5\Phi)\mu_w \tag{1}$$

Density is calculated using Pak and Cho equation which is given as follows:

$$\rho_{nf} = v\rho s + (1 - \vartheta)\rho_w \tag{2}$$

Nanoparticles taken for heat exchanger modelling in this work are in very low concentration (up to 0.025%). Therefore, above equation can be used for calculating the density. Specific heat for the nanofluid can be calculated using the equation given below [10].

$$C_{pnf} = \frac{\Phi(\rho_s . C_{ps}) + (1-\Phi)(\rho_w . C_{pw})}{\rho_{nf}} \tag{3}$$

Suspension of higher thermal conductivity nanoparticles in fluid have profound impact on the thermal conductivity of fluid. There are many models existing in the literature providing the enhancement in the thermal conductivity after the preparation of nanofluid. Many experiments were performed and then compared with the analytical results. There is further scope of research in this but there are some empirical correlations available in the literature which effectively calculate thermal conductivity of the binary phase mixture.

In the literature available, the thermal conductivity has been calculated with the help of thermal conductivity enhancement ratio which can be defined as the ratio of Thermal conductivity of nanofluid to that of base fluid (K_{nf}/K_l). The effective thermal conductivity of nanofluid was calculated, using the correlation introduced by Hamilton-Crosser [11], as follows

$$K_{nf} = K_f \frac{(Kp+2K_f+2\Phi(Kp - K_w)}{Kp+2K_f - \Phi(Kp - K_w)} \quad (4)$$

The equation is valid for binary phase mixture with dilute suspension velocity. Equation 4 can be applied in this case to get the reasonable approximate estimation. Pak and Cho examined correlation of heat transfer in turbulent flow using experimental results [12]:

$$N_{unf} = \frac{h_{nfd}}{K_{nf}} = 0.021({Re_{nf}}^{0.8})({Pr_{nf}}^{0.5}) \quad (5)$$

The Reynolds number, Peclet number and Prandtl number are defined respectively as [13]:

$$Re_{nf} = \frac{\rho_{nf} VaD}{\mu_{nf}} \quad (6)$$

$$\text{Pec} = \frac{VavDiap}{\propto nf} \quad (7)$$

And

$$Pr_{nf} = \frac{\mu_{nf} C_{pnf}}{K_{nf}} \quad (8)$$

Whereas, the thermal diffusivity for nanofluid is given by,

$$\propto_{nf} = \frac{K_{nf}}{(\rho Cp)nf} \quad (9)$$

The Peclet number is used to get an idea of thermal dispersion caused by suspended particles microdiffusion. Now similar to heat transfer coefficient, the friction factor is defined as [4]:

$$f_{nf} = \frac{2D\Delta p_{nf}}{L\rho_{nf} u_m^2} \quad (10)$$

This work considers two dimensional long enclosure in which fluid is flowing at steady state. The horizontal walls are conducting heat and are impermeable to mass transfer. The nanofluid flowing inside the enclosure is Newtonian and incompressible. The flow is turbulent and it is assumed that the nanoparticles are flowing at the same velocity as that of fluid. Moreover, it is assumed that nanoparticles have uniform shape and size. The boundary conditions (B.C.) with these assumptions in this particular investigation, are presented as

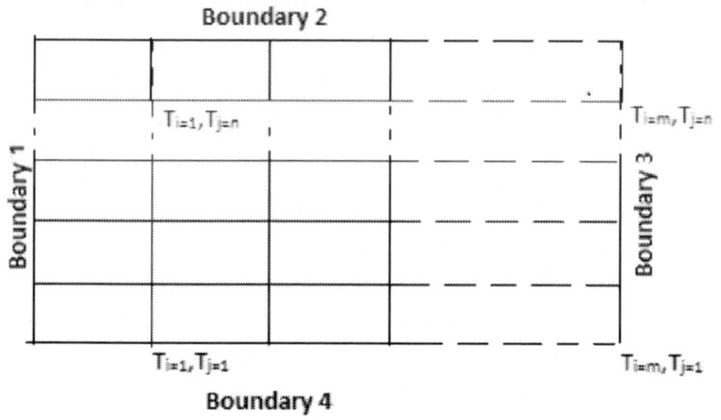

Fig. 1- Two dimensional rectangular enclosure

Considering a rectangular enclosure (ref. fig. 1) divided into rectangular mesh of size 100 x 100. The mesh was made more precise in the radial direction since the effect in radial direction is more important and the properties are varying significantly along that direction. Flow at uniform temperature of 313K is entering the tube and shell side fluid is following linear variation of temperature. The linear variation of temperature effluent wall is considered same as that of the inside wall.

$$T_{hotin} = 363\ K \quad \text{(B.C. 11.1)}$$

$$T_{hotout} = 333\ K$$

Thus boundary 1 follows the equation 2 for the temperature variation. The hot temperature inlet is at 363 K and effluent is coming at 333 K

$$T = \frac{(T_{hotin} - T_{hotout})}{l} \times \left(\frac{z}{Nx}\right) \times \frac{i}{Nx} + T_{hotout} \quad \text{(B.C. 11.2)}$$

Boundary 3 is assumed to be isothermal that is temperature is same as that of the effluent temperature of tube side.

$$T_{i=mj=1:n} = T_{coldout} \quad \text{(B.C. 11.3)}$$

Due to symmetry, temperatures are similar to the other half of the tube therefore, boundary 4 follows:

$$T_{j=1::m,j=1:n} = \text{same as the other half of the tube} \quad \text{(B.C. 11.4)}$$

The flow in heat exchanger is generally turbulent therefore, the Reynolds number is varied from 3000 to 40000. Temperature range for the cold fluid is in the range of 20-70 degree Celsius. The cold fluid or the nanofluid is flowing inside the inner tube and hot fluid is flowing in the outer shell. Mean value of parameters such as density, viscosity etc., in the given range are calculated and used further in the calculations. These properties of nanofluids vary significantly with materials. The Thermal and Physical properties for all fluids are presented in Table 1. It is considered that convective momentum is playing major role in the axial direction and shear stress effects has its major role in the radial direction (Vr, Vϴ=0). The hydrodynamic boundary layer inside the tube is neglected. Considering the assumptions governing equation was formulated and was discretized using finite volume approach. Governing equations were first simplified to the general differential equations as follows:

Shear Stress across radial direction

$$\tau_{rz} = -\mu\left[\frac{\partial V_r}{\partial z} + \frac{\partial V_z}{\partial r}\right] \quad (12)$$

Shear Stress across angular direction

$$\tau_{\theta z} = -\mu\left[\frac{1}{r}\frac{\partial V_z}{\partial \theta} + \frac{\partial V_\theta}{\partial z}\right] \quad (13)$$

Shear Stress across axial direction

$$\tau_{zz} = -\mu[2\frac{\partial Vz}{\partial z} + \mu\frac{\partial Vz}{\partial z}] \quad)14($$

$$\mathrm{Vz}\frac{\partial Vz}{\partial z} = -\frac{\partial P}{\partial z} + \frac{\mu}{r}\frac{\partial Vz}{\partial r} + \mu\frac{\partial 2Vz}{\partial r^2} + \mu\frac{\partial 2Vz}{\partial r^2} \quad (15)$$

While calculating temperature using energy balance, the viscous heat effects have been considered to be negligible. Convection in axial direction and conduction in radial direction were assumed. Flow was modelled as counter current in which the temperature distribution on shell side inner surface is assumed similar to inside surface of tube. Similar to that of momentum the thermal boundary layer is also neglected. Considering the assumptions, the governing equation for energy is formulated as follows:

$$\rho C_p\left[V_z\frac{\partial T}{\partial Z}\right] = \frac{1}{r}\left[K\frac{dT}{dr}\right] + K\frac{d^2T}{dr^2} + K\frac{d^2T}{dz^2} \quad (16)$$

Table-1 Thermal and physical properties for fluids.

Nano fluid or base fluid	**Concentration volume %**	**Density (Kg/m^3)**	**Specific heat (KJ/kg)**	**Conductivity (W/m.K)**	**Reference**
Water	0	985.7	4179	0.623	14
Graphene/water	0.1, 0.075, 0.05, 0.025	985.7-994.9	4150.5-4179	0.62-0.817	14
Al_2O_3	0.1, 0.054, 0.03, 0.025	3700	880	46	15
TiO_2	0.1, 0.05, 0.025, 0.01	4170	711	11.8	16

Results and Discussion

In the present investigation, the model was simulated using MATLAB for wide range of operating parameters. The mesh size of 100 x 100 was maintained and the heat transfer enhancement along with the undesirable effect of pressure drop were calculated.

Temperature and pressure distribution:

The temperature profile along the length of tube is shown in fig. 2a. It shows the temperature distribution in radial direction. Temperature profile along the length of tube is approximately linear therefore, temperature is proportionally increasing with the length. In radial direction conductive heat transfer following second order differential equation is more significant, therefore the variation is nonlinear (ref. fig. 2.b). Along the length, convective heat transfer coefficient is more dominating and variation is therefore, linear.

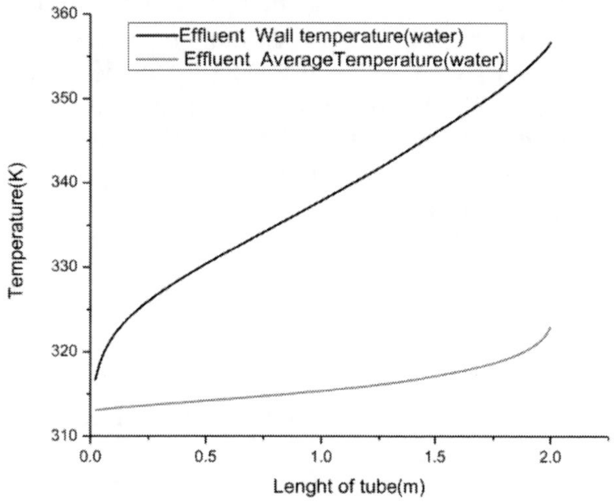

Fig 2.a- Axial variation of temperature

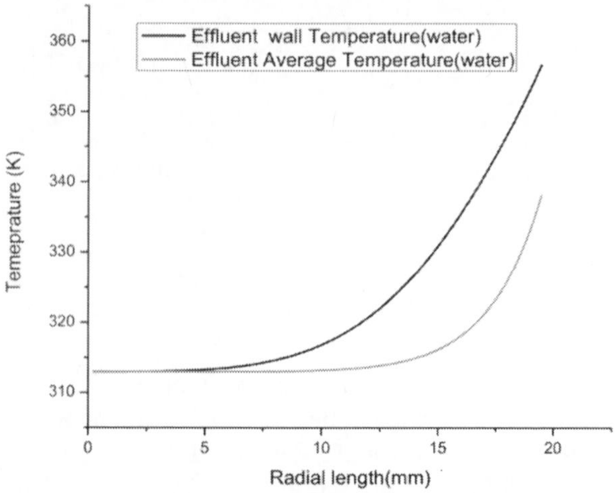

Fig 2.b Radial variation of temperature

Reynolds number was varied in the normal range of 3000-40000 and the pressure drop seems to increase with the Reynolds number. The increase is notably more at high Reynolds number where the exponential rise of pressure drop can be seen. The change can be observed due to the resistance developed in the natural flow by the nanofluid particles. The erratic movement of nanoparticles dispersed in fluid collides with fluid particles which leads to energy loss and results in increase of pressure drop (ref. fig. 3).

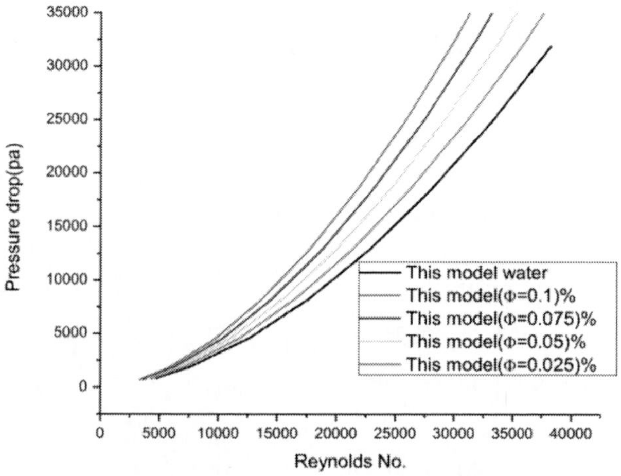

Fig. 3- Comparison of Pressure drop and Reynolds number at different concentrations of graphene nanoparticles (0.1 %, 0.075 %, 0.05 %, 0.025 %)

Thermal Energy Transfer

Thermal conductivity which was calculated using equation 4 varies from 0.62 W/m^2K for base fluid to 0.82 W/m^2K for the concentration of 0.1% of nanoparticle. Maximum of 32 % enhancement was observed using nanofluid of this concentration. High thermal conductivity and high Reynolds number result in better heat transfer rate. The fig. 4 shows this increase in convective heat transfer coefficient. Up to 600% augmentation in the heat transfer rate is observed at maximum turbulence and 0.1 % concentration of nanoparticles when compared with lowest Reynolds number of base fluid as a reference. When compared at same Reynolds number the change observed in convective heat transfer coefficient above base fluid found to be in the range of 2 % - 300 %. This shows that the fluid with large momentum results in greater increase in heat transfer coefficient. This may be due to the dispersion of nanoparticle with better thermal conductivity, in nanofluids because as the concentration of nanoparticle increases its movement along the fluid drives the heat at higher rate. Near the wall where the velocity of fluid is about zero, the conductive heat transfer coefficient for fluid is more dominating, whereas, at the middle of the tube its effect is marginal.

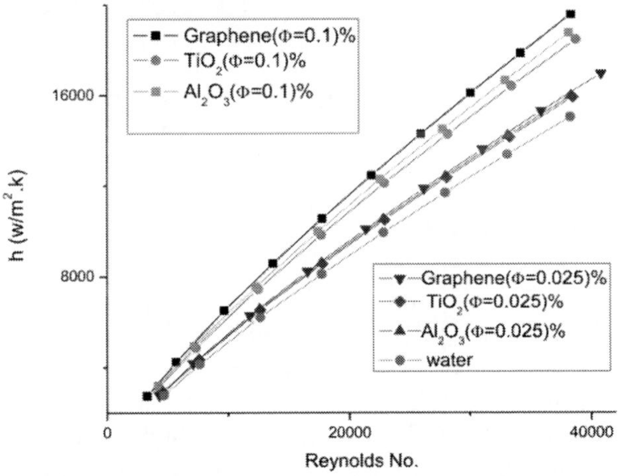

Fig. 4- Comparison of thermal performance for nanofluids at different concentrations

Fig. 4 shows the variation of convective heat transfer coefficient for different nanofluids at concentration of 0.1 % and 0.025 %, with Reynolds number ranging from 3000-40000. Graphene particles with concentration of 0.1 % (by volume), shows the highest thermal performance among all the fluids considered in this study. At low concentrations, the thermal performance of different nanofluids does not vary much as compare to that at higher concentrations. Also the other advantages of using graphene as nanofluid are that along with than thermal conductivity, the thermal and physical properties like density, specific heat of it are similar to base fluid water whereas, in case of Al_2O_3 and TiO_2, there is a significant difference in the properties given in Table 1. On its counterpart, pressure drop is almost equal for same concentrations of different nanofluid. Thus, graphene/water nanofluid utilization gives better thermal performance over other nanofluids like TiO_2/water and Al_2O_3/water etc. The maximum and minimum enhancement in the thermal performance of graphene/water nanofluid at concentrations of 0.1 % is found to be 300 % and 40 % in the Reynolds number range of 3000-40000. For the concentration of 0.025 % maximum enhancement is found to be 57% at low Reynolds number range and minimum enhancement found to be 9% at high Reynolds number. In case of TiO_2/water, at concentration of 0.1 %, about 25 % enhancement is found. At 0.025% concentration the rise of 5% in thermal performance with respect to base fluid can be observed. In case of Al_2O_3/water, at concentration of 0.1% and 0.025%, enhancement of 27% and 6% is found respectively

Energy Factor

Response of nanofluids in terms of thermal energy factor and pressure energy factor is evaluated and its relative comparison is shown in fig 5. Energy factor are calculated for different concentrations of nanofluid with reference to the base fluid. Utilization of nanofluids leads to augmentation in thermal energy transfer but also causes pressure drop. As the Reynolds number increases the thermal energy factor decreases and at high Reynolds number thermal energy factor becomes constant. Thermal energy factor gives the thermal performance of nanofluid with respect to base fluid. Therefore, greater the thermal energy factor, higher the thermal performance of nanofluid. Similarly pressure energy factor gives the pressure drop of nanofluid to that of base fluid therefore, it should be near to 1 to corroborate that nanofluid has better overall performance. With Reynolds number the pressure energy factor start decreasing in the beginning then it increases in the between. Finally at higher Reynolds Number Pressure energy is becoming constant for less concentrated nanofluid (less than 0.1 % concentration). Comparison of both energy factors can help to decide the velocity of nanofluid that can be optimal for the performance of heat exchanger.

$$Thermal\ energy\ factor = \frac{h_{nf}}{h_b} \quad (17)$$

$$Pressure\ energy\ factor = \frac{\partial P_{nf}}{\partial P_b} \quad (18)$$

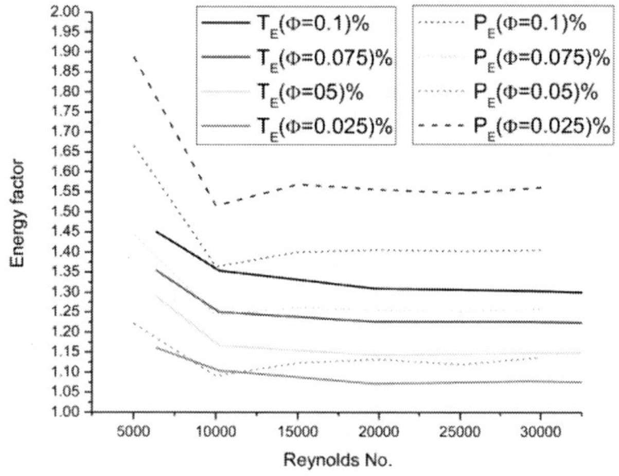

Fig . 5 Energy factor for different concentration

Pressure energy factor is observed to be higher as compared to the thermal energy factor at low concentrations. As concentration is increased, this gap increases. At lower concentration (0.025 %) thermal energy factor varies in the range of 1.15 to 1.10 whereas, the pressure energy factor varies in between 1.2 and 1.5, but in case of higher concentration (0.1 %), thermal energy factor varies from 1.45 to 1.3 and pressure energy factor lies in between 1.9 and 1.6. At high Reynolds number both energy factors are constant, this may be due to the less effect of augmentation of nanoparticle. At high velocity, the effectiveness of nanoparticle becomes constant, as a result nanoparticle potential does not change much in highly turbulent regions. This is because of the reduction in particle residence time in the tube. The comparison is made for heat transfer coefficient with respect to change in pressure drop for different concentration (ref. fig. 6). It can be observed that the slope decreases with increase in pressure this shows that with increase in Reynolds number pressure drop increases and relative effect of thermal performance decreases. For higher concentration, although thermal performance is high, it also has significant pressure drop.

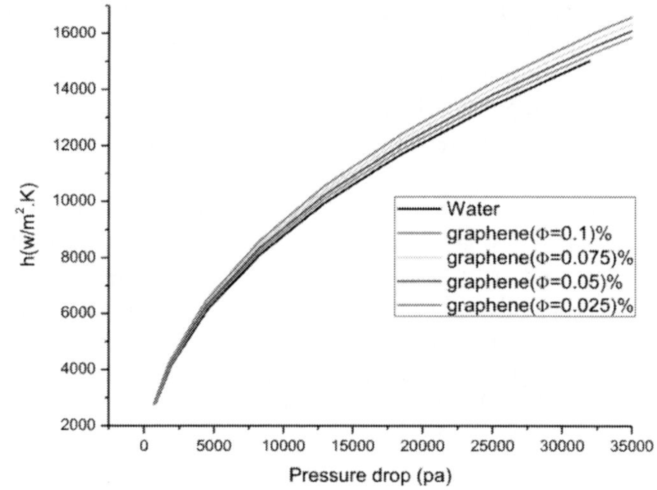

Fig. 6- Variation of convective heat transfer with pressure drop at different concentrations of nanofluid

Experimental validation

Simulation parameters are taken from literature for all three types of nanofluids. L/D ratio is considered greater than 10 in all cases. The accuracy of this model in terms of the percentage error is shown in Table 2, in which correlation factor R^2 along with the average percentage error are calculated for nonlinear system and listed [17]. Comparison of this model with the experiments performed for Graphene-water nanofluids in the heat exchanger system using concentration of 0.1

%, 0.075 %, 0.05 %, 0.025 % (ref. fig. 9). Fig. 7 shows the experimental data comparison with the model for friction factor using base fluid. Comparative study for heat transfer coefficient with experimental results of Ranjbarzadeh et al., is shown in fig. 9 [14].

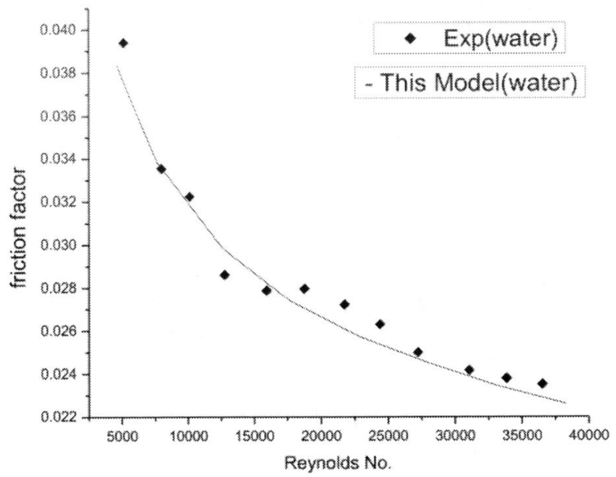

Fig. 7- Comparison of friction factor with experimental results [14]

Model is verified by comparing it with experimental results for ɣ-Al_2O_3/water nanofluid in fig. 10 where concentrations of nanoparticles were 0.03 % and 0.054 % (volumetric basis) [15]. The

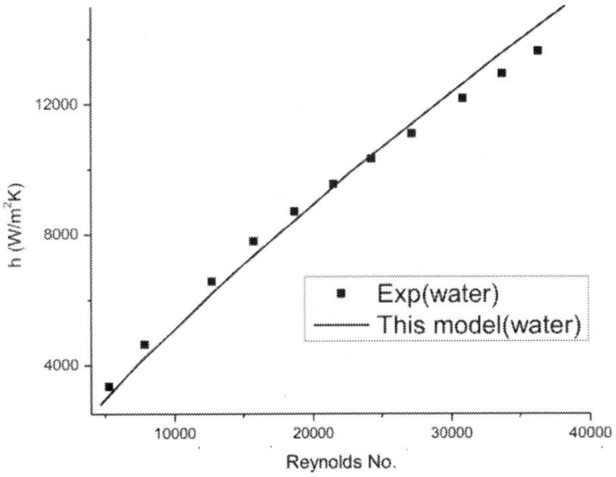

Fig. 8- Comparison of convective heat transfer coefficient with experimental results [14]

accuracy of data with these experimental results were found to be within the range of 0 % - 8 % at different data points. Similar results were obtained in case of TiO_2/water nanofluid [16], where two concentrations 0.1 % and 0.05 % (by volume) are compared and R^2 value obtained in range of 97 -99 . The percent error obtained in the range of ±3% as shown in Table 2.

Table 2- Comparison of the simulation results for different nanofluids with experimental data from the literature [14-16].

Fluid	Property	R^2	Average % error
Water	Friction factor	99.9	4.16
Water	h (W/m^2K)	98.8	1.08
Graphene (0.1%)	h (W/m^2K)	98.2	3.54
Graphene (0.075)%	h (W/m^2K)	98.4	3.54
Graphene (0.05)%	h (W/m^2K)	98.5	3.21
Graphene (0.025)%	h (W/m^2K)	98.9	5.77
TiO_2 (0.05)%	h (W/m^2K)	97.5	-3.19
TiO_2 (0.1)%	h (W/m^2K)	98.8	-2.39
Al_2O_3 (0.03)%	h (W/m^2K)	98.5	7.35
Al_2O_3 (0.054)%	h (W/m^2K)	99.7	5.93

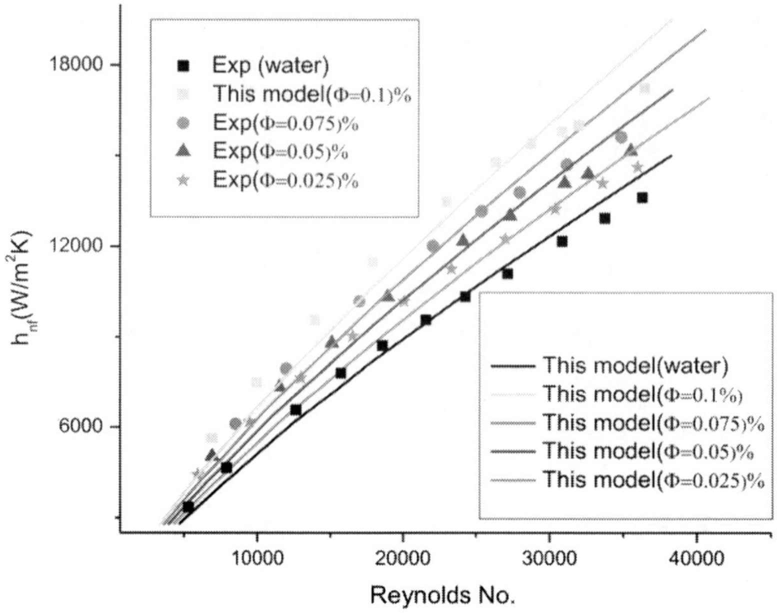

Fig. 9- Comparison of heat transfer coefficient with the experimental results [14]

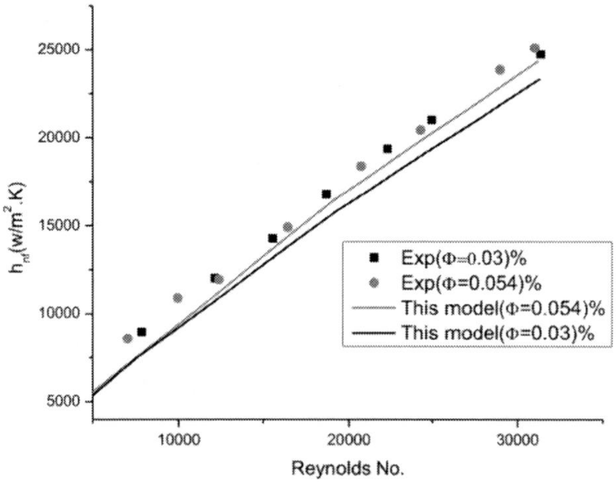

Fig. 10- Comparison of Al_2O_3/water nanofluids with experimental results [15]

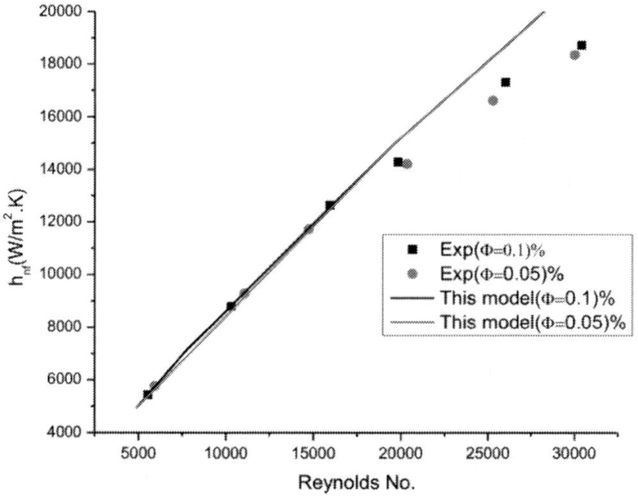

Fig. 11- Comparison of TiO_2/water nanofluid with experimental results [16]

Conclusions

Based on this study, following conclusions are drawn:

- With increase in Reynolds number and concentration of nanoparticles, friction factor of the nanofluid decreases and the total pressure drop increases.
- The effect of convective heat transfer in axial direction is more and in radial direction effect of conductive heat transfer is more significant
- Pressure energy factor is more dominating over thermal energy factor at higher concentrations. But the net increase in thermal performance coefficient due to nanoparticle is way more than the reference base fluid 2 %-300 %. Whereas, the maximum pressure drop found is 10%-88% for same Reynolds number. Therefore, the thermal performance can dominate over the increase in pressure drop for some Reynolds number in turbulent region.
- The model is found to be well fitted for nanofluid flow in turbulent region at lower concentrations, with an accuracy of R square value in the range of 97% - 99% with the maximum percentage error of 7.
- The wall temperature variation is high as compared to bulk fluid because of augmentation of nanoparticles near the tube wall.
- Utilization of nanofluid in heat exchange medium is found to be beneficial in terms of net performance of heat exchanger as the energy factor obtained is 1.6, which ensures quantitatively better heat transfer improvement over pressure drop.
- Utilization of graphene/water nanofluid in heat exchanger promises better thermal performance over other nanofluid TiO_2/water and Al_2O_3/water with enhancement varying in range of 40%-300% and 5 % - 8 % for concentration of 0.1 % and 0.025 % respectively.

Acknowledgement

Authors wish to extremely thank Mr. Amit Kumar for his timely and valuable inputs throughout the work. Authors are also thankful to Birla Institute of Technology and Science-Pilani, Rajasthan for providing the facilities to carry out this research work.

References

[1] Raj Chengappa (2015), "Why the world can't save the earth," India Today, December, pp. 10-15

[2] Bond, G.M., Stringer, J., Brandvold, D.K., Simsek, F.A., Egeland, G. (2001), "Development of Integrated System for Biomimetic CO_2 Sequestration Using the enzyme Carbonic Anhydrase," Energy. fuel. 15, 309–316.

[3] Thomas, D.M., Mechery, J. and Paulose, S.V. (2016), "Carbon dioxide capture strategies from flue gas using microalgae: a review," Environmental Science and Pollution Research, 23(17), pp. 16926-16940.

[4] Duangthongsuk, W. and Wongwises, S. (2009), "Heat transfer enhancement and pressure drop characteristics of TiO_2–water nanofluid in a double-tube counter flow heat exchanger," International Journal of Heat and Mass Transfer, 52(7-8), pp.2059-2067.

[5] Kumar, P.M., Kumar, J. and Suresh, S. (2013), "Experimental investigation on convective heat transfer and friction factor in a helically coiled tube with Al_2O_3/water nanofluid," Journal of Mechanical Science and Technology, 27(1), pp.239-245.

[6] Assael, M., Chen, C.F., Metaxa, I. and Wakeham, W.A. (2004). "Thermal conductivity of suspensions of carbon nanotubes in water," International Journal of Thermophysics, 25(4), pp.971-985.

[7] Fotukian, S.M. and Esfahany, M.N. (2010), "Experimental study of turbulent convective heat transfer and pressure drop of dilute CuO/water nanofluid inside a circular tube," International Communications in Heat and Mass Transfer, 37(2), pp.214-219.

[8] Soldano, C., Mahmood, A., Dujardin, E., (2010), "Production, properties and potential of graphene,". Carbon. carbon.2010.01.058

[9] Drew, D.A. and Passman, S.L. (2006), "Theory of multicomponent fluids," Springer Science & Business Media.

[10] Prabhat, N., Buongiorno, J. and Hu, L.W. (2011), "Convective heat transfer enhancement in nanofluids: real anomaly or analysis artifact," In ASME/ JSME 2011 8th Thermal Engineering Joint Conference, American Society of Mechanical Engineers, January, pp. T10070-T10070.

[11] H.L, Hamilton O.K. Crosser, Thermal conductivity of heterogeneous two component systems, I and EC Fundam. 1 (3) (1962) 187–191.

[12] B. C. Pak and Y. I. Cho, "Hydrodynamic and heat transfer study of dispersed fluids with submicron metallic oxide particles," Experimental Heat Transfer, vol. 11, no. 2, pp. 151–170, 1998

[13] Kakaç, S., Pramuanjaroenkij, A. (2009), "Review of convective heat transfer enhancement with nanofluids", International Journal of Heat and Mass Transfer, 52, pp. 3187–3196.

[14] Ranjbarzadeh, R., Karimipour, A., Afrand, M., Isfahani, A.H.M. and Shirneshan, A., (2017) "Empirical analysis of heat transfer and friction factor of water/graphene oxide nanofluid flow in turbulent regime through an isothermal pipe," Applied Thermal Engineering, 126, pp.538-547.

[15] Fotukian, S.M. and Esfahany, M.N., (2010), "Experimental investigation of turbulent convective heat transfer of dilute γ-Al2O3/water nanofluid inside a circular tube," International Journal of Heat and Fluid Flow, 31(4), pp.606-612.

[16] Sajadi, A.R. and Kazemi, M.H., (2011), "Investigation of turbulent convective heat transfer and pressure drop of TiO2/water nanofluid in circular tube," International Communications in Heat and Mass Transfer, 38(10), pp.1474-1478.

[17] Brown, A.M., (2001) "A step-by-step guide to non-linear regression analysis of experimental data using a Microsoft Excel spreadsheet," Computer methods and programs in biomedicine, 65(3), pp.191-200

532284 Electrochemical Studies of Low Temperature Ionic Liquid-Cosolvent-Salt Electrolyte Systems

Abstract:

Ionic liquids (IL), consist of ions and remain as liquid at ambient temperatures, have shown potentials as electrolytes in a great variety of applications. Their characteristic traits (low phase transition temperature, low volatility, good stability, etc.) broaden the operational temperature range on both high and low ends, while incorporations with other solvents and salts appear to be a common strategy for optimizations and enhancements on both physicochemical and electrical properties. Understandings in electrochemical properties of the IL-cosolvent-salt systems, developed for molecular electronic transducer (MET) seismometer to conduct seismology studies under extreme conditions, are crucial for evaluations of the functionality and future improvements. In this work, we present new findings in electrochemical aspects relevant to performances of the MET seismometer, specifically on the ionic conductivity and electrochemical stability. The electrochemical impedance spectroscopy technique was employed to measure the conductivity evolution from room temperature to around -70 ËšC. The results revealed that the ionic conductivity resembled an Arrhenius behavior, and did not show significant breakdowns before reaching the glass transition temperature (Tg). Cyclic voltammetry was applied to study the electrochemical windows, with platinum as the working electrode. Within an appropriate potential window, iodide (I^-)/triiodide (I_3^-) redox species reacted and generated currents without additional undesired electrochemical reactions occurred. The current best developed IL-water-salt electrolyte system demonstrates a desirable conductivity ranging from room temperature to -70 ËšC, and a good electrochemical stability under an properly applied voltage. In addition to electrochemistry investigations, results from computational works provide insights of interactions on the molecular level and possible explanations for observed phenomena. Discoveries of the presented works may serve as a reference for MET seismometer applications, and other low temperature electrochemical devices operating based on the I^-/I_3^- redox pairs.

Stability of Oil-in-Water Emulsions of Heavy Crude under Flowing Conditions

Yue Cui, China University of Petroleum-Beijing, Beijing, China
Qiyu Huang, China University of Petroleum-Beijing, Beijing, China
Jiadi Zhao, China Huanqiu Contracting & Engineering Co., Ltd., Beijing, China,
Weidong Li, China University of Petroleum-Beijing, Beijing, China
Caoding Wang, China University of Petroleum-Beijing, Beijing, China

Abstract: A unique experimental device was constructed to investigate the emulsion stability of heavy crude. It mainly consists of a stirred tank with good temperature control, a stirring system equipped with a torque meter and a V-shaped measuring tube. Sensitivity study regarding temperature, salinity, stirring speed, emulsifier dosage, initial water fraction of oil-water mixture was executed by measuring the free water after stirring. In this study, dynamic water fraction was proposed as free water fraction under flowing condition. It was found that the dynamic water fraction of heavy crude decreases with salinity and emulsifier dosage, and increases with temperature. Based on the experimental results, we initially established a water separating fraction model for heavy crude, applicable in both static and flowing conditions. Verification shows a good agreement between prediction and experiment with average relative error below 23.72%. Calculating water separating fraction, this model benefits in estimating pipeline flow condition, and further helps to schedule transportation arrangements.
Keywords: Interfacial Phenomena, Transport Phenomena, Measurement

1 Introduction

Oil-water and oil-gas-water multiphase flow widely exist in crude oil production and transportation. When the fluid flows through the pipes, the elbows, the valves and other accessories, the shearing effect triggers formation of emulsions, especially at the existence of interfacial active substances such as asphaltene, colloidal and waxy crystal. The emulsions change the proportion of oil-water phase and enhance the effective viscosity of the fluid, thus further affect the flow pattern and pressure drop [1-3].

Under static conditions, free water forms from the oil-water system over time. But when the system is continuously sheared, there exists a dynamic balance between rupture and coalesce of the emulsion droplets, thus the emulsion has a certain water-containing stability. By far, static stability of emulsion has received wide attention, while the dynamic stability however is way less focused.

The stability of crude oil emulsion mainly depends on the oil-water interface property. The interfacial active components can be adsorbed on the oil-water interface, forming a viscoelastic interface film with certain strength, which act as barriers for droplet coalescence. In this work, emulsion stability is characterized by dynamic water fraction.

The emulsification of oil-water system is not only a strong function of oil and water properties such as salinity, emulsifier dosage, pH value, it is also affected by external factors including comprehensive water content, stirring rate and temperature [4-9]. In response, this paper aims to investigate the stability of oil-in-water emulsions of heavy crude under flowing conditions. The effects of salinity, emulsifier dosage, total water fraction, stirring speed and settling temperature on emulsion stability were experimentally studied. An empirical model for predicting the dynamic water fraction was accordingly proposed.

2 Experimental Section

2.1 Materials

Oil and water samples were collected from Bohai sea oilfield in China. An overview of some physical properties are listed in Table 1, Table 2 and Figure 1, respectively.

Table 1. Physical Properties of Oil

Oil Sample No	Pour Point(°C)	Density at 20 °C (kg/m^3)
1	56	945.6
2	21	990.3

Table 2. Composition of Water Sample

Ion	Mg^{2+}	Ca^{2+}	HCO_{3-}	Na^{+}+K^{+}	CO_3^{2-}	Cl^-	SO_4^{2-}
Salinity (mg/L)	128.17	212.00	1507.19	3766.00	675.00	6000.00	288.10

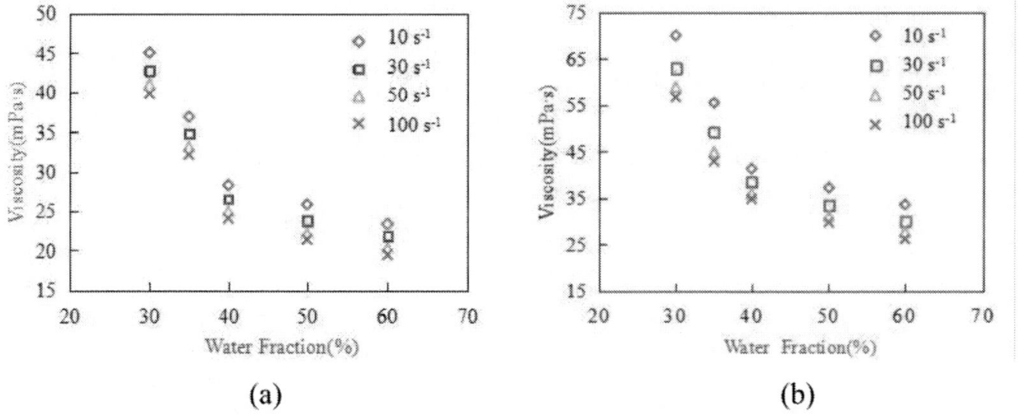

Figure 1. Viscosity Curves of Emulsions for Oil Samples 1 (a) and 2 (b)

2.2 Experimental Apparatus

As shown in Figure 2, the experimental apparatus mainly consists of: (1) a 75-mm-ID and 105-mm-length oil tank with a bottom discharge valve. It is a bilayer structure collected to a thermal water bath for temperature control; (2) an electric overhead stirrer equipped with an impeller and a tachometer. The effective stirring speed ranges from 10 to 1600 rpm; and (3) a V-shaped measuring tube to determine the volume of the separated water from the oil-water mixture.

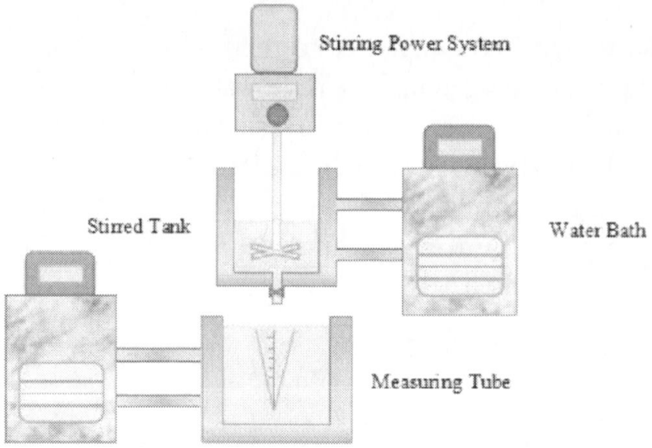

Figure 2. Sketch of the Experimental Apparatus

2.3 Test Procedures

(1) Initially, preheat the oil tank to the desired temperature. Then load the crude oil and pure water with emulsifier into the tank separately and hold for 10 min. Their volumes were carefully calculated according to the experimental requirement.

(2) Start the stirrer at a designated and constant speed for 10 min to achieve equilibrium mixing. The torque of the stirrer was automatically recorded. The stirring speed ranges from 200 rpm to 1000 rpm.

(3) After stirring, open the bottom valve to transfer the oil-water mixture into V-shaped tube. Then record the volume of free water after different static time. Particularly, each experiment was repeated three times, and the mean values were taken as the final result.

3 Results and Discussion

3.1 Determination of Dynamic Free Water Fraction

Figure 3 illustrates the variation of free water over time.The formula of free water fraction is calculated as follows:

$$\phi_{w} = \frac{V_{f}}{V_{w}} \times 100\% \tag{1}$$

Where V_f is the volume of free water at different settling time in milliliters, V_w is the volume of water in the mixture in milliliters.

It was found that the free water fraction logarithmically grows with settling time (see Figure 4). Thus the relationship between free water fraction and time can be written as

$$V_{f} = a\ln(t+b) + c \tag{2}$$

Where V_f is the volume of free water at different settling time in milliliters; t is settling time; *a,b* and *c* are undetermined coefficients varied with component of oil-water system and emulsification conditions.

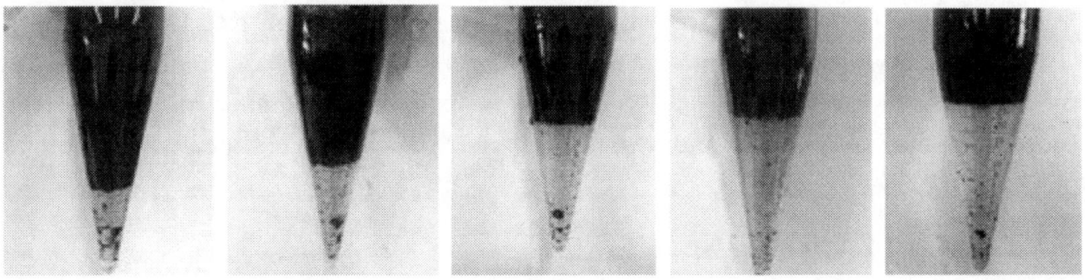

10s 30s 60s 120s 180s

Figure 3. Accumulation of Free Water

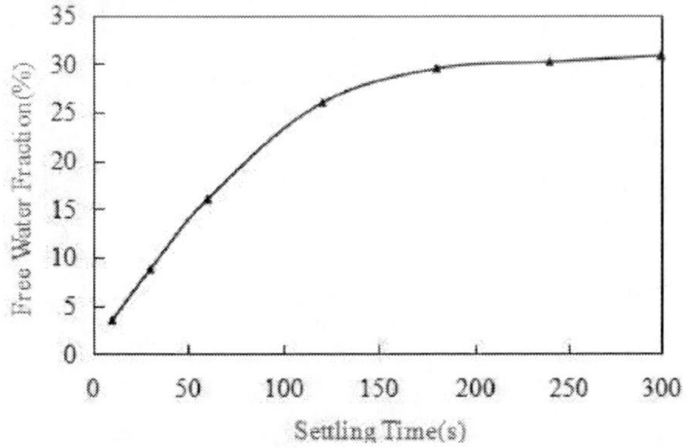

Figure 4. Curve of Settling Time vs Separated Water of Crude Oil

In this study, the free water fraction under flowing condition, named as the dynamic water fraction, is calculated by the equation above with t=0. When the volume of free water in the V-shaped measuring tube is equal to 0, it is defined as the full emulsification of water, and if not, it is defined as the partial emulsification of water.

3.2 Sensitive Analysis

3.2.1 Effect of Initial Water Fraction

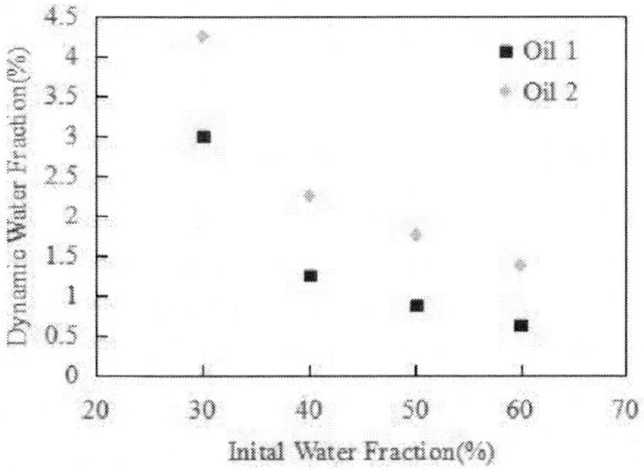

Figure 5. Effect of Water Fraction on Dynamic Water Fraction (Emulsifier Dosage = 500ppm, Stirring Speed = 500r/min, Stirring Time = 10min, pH = 7)

Figure 5 shows the variations of dynamic water fractions of oil 1and oil 2. The test matrix include 500 ppm of emulsifier dosage, 500 r/min of stirring speed, 10 min of stirring duration and the pH value was 7. It can be seen that increase of initial water fraction promotes the emulsion stability with a turning point of 40%. Two oil sample is more stable when water fraction is over 40%. When the water content is too low, the stable emulsion cannot be formed.This is because higher dispersed phase proportion makes oil drop and water drop more easily to collide, which helps to produce stable O/W emulsion.

3.2.2 Effect of Stirring Speed

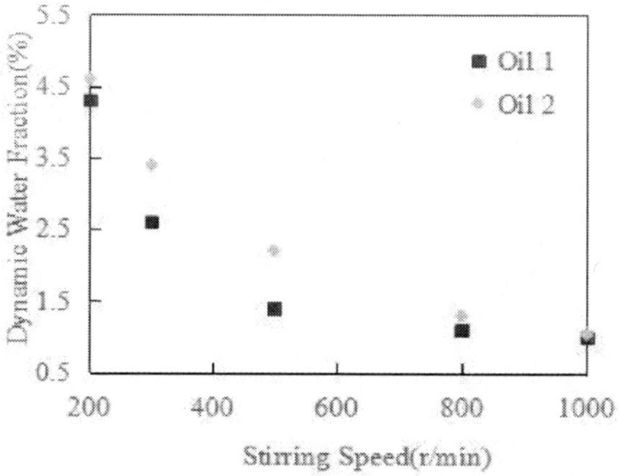

Figure 6. Effect of Stirring Speed on Dynamic Water Fraction(Emulsifier Dosage = 1000ppm, Water Fraction = 60%, Stirring Time = 10min, pH = 7)

The effect of stirring speed on dynamic water fraction is given in Figure 6, experiments were performed for systems of the two oils that varies with the thermostatic time with different stirring speed (r) at 50°**C**. The emulsifier dosage was 1000ppm, the water fraction was 60%, the stirring time was 10min and the pH value was 7. It can be seen that when the mixture are stirred at a low speed, only part of the water could be emulsified. High stirring speed triggers full emulsification, where the emulsified water fraction being equal to the water fraction of the system. This is because high stirring speed decreases the grain diameter of oil drop, thus the interfacial tension of oil-water interface gets enhanced and emulsion stability accordingly rises.

3.2.3 Effect of Temperature on Settling Range

The variations of dynamic water fraction at different temperatures is given in Figure 7. Increasing settling temperature generally elicits higher water fraction. It is attributed to the increased droplet collision and water/oil density difference at high temperatures. Aggregation, coalescence and sedimentation of droplets accordingly get enhanced. Thus more separated water forms.

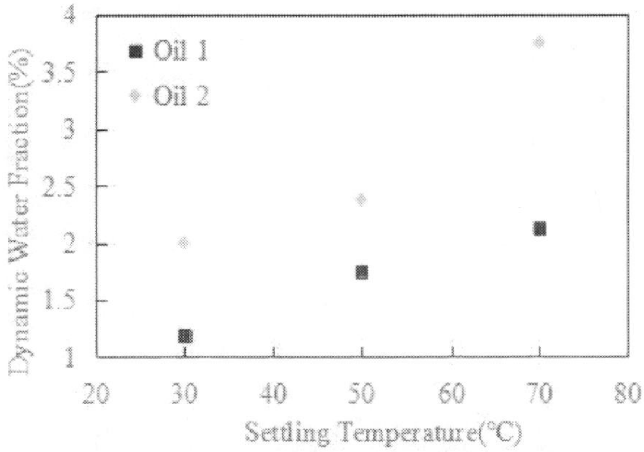

Figure 7. Effect of Stationary Temperature on Dynamic Water Fraction(Emulsifier Dosage = 500ppm, Water Fraction = 60%, Prepare Temperature = 80 °C, Stirring Time = 10min, pH = 7)

3.2.4 Effect of Salinity

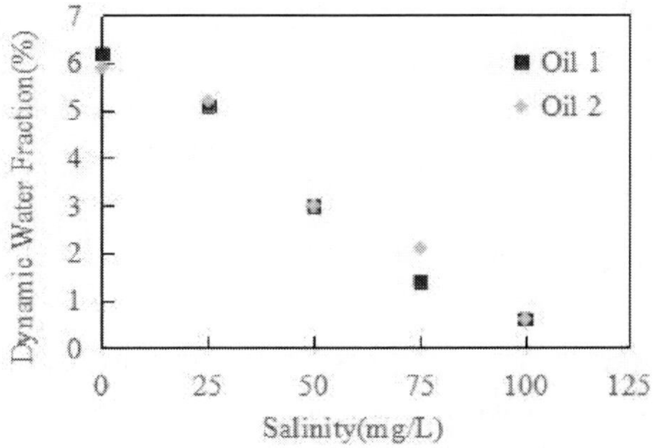

Figure 8. Effect of Salinity on Dynamic Water Fraction(Emulsifier Dosage = 1000ppm, Water Fraction = 60%, Prepare Temperature = 50 °C, Stirring Speed = 500r/min, Stirring Time = 10min, pH = 7)

As shown in Figure 8, increasing water salinity reduces the dynamic water fraction of the oil-water system and improves the stability of emulsion. This is ascribed to the fact that the inorganic salt in water weakens the repellent force at oil-water interface of ionic surfactant in the oil, thus more interface active agent can be absorbed to form a stronger barrier between oil and water, which decreases the oil-water interfacial tension.

3.2.5 Effect of Emulsifier Dosage

As shown in Figure 9, the dynamic water fraction of O/W emulsion at the same time decreased significantly with the increase of emulsifier dosage, and the stability of emulsion enhanced obviously for the same oil.

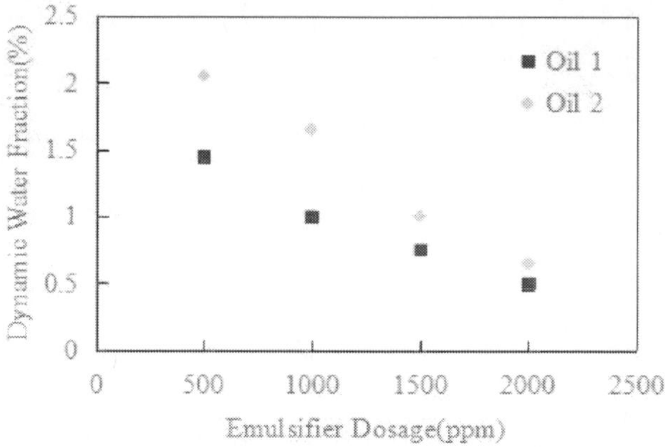

Figure 9. Effect of Emulsifier Dosage on Dynamic Water Fraction (Emulsifier Dosage = 1000ppm, Water Fraction = 60%, Prepare Temperature = 50 °C,Stirring Speed = 500r/min, Stirring Time = 10min, pH = 7)

O/W emulsion is a heterogeneous dispersion system with continuous phase interface and large interfacial energy between the oil drops (dispersed phase) and water drops (continuous phase). The oil-water system tends to reduce the interfacial energy spontaneously, which causes coalescence of drops. The O/W emulsion is a thermodynamically unstable system. Emulsion decay is highly correlated the flocculation and coalescence of water and oil drops. It is a relatively dynamic process.

4 Modeling Dynamic Water Fraction

4.1 Model Development

The above discussions have proved that the dynamic water fraction under flowing condition decreases with water cut, emulsifier dosage and stirring speed and increases with temperature. Based on this, a model for predicting the water fraction under flowing condition is established

$$\phi_w = k\exp(a\phi + b/r + c\omega + dT)\ln(mt + n) \tag{3}$$

Where ϕ_w is the water syneresis rate under flowing condition; ϕ is the initial water fraction; r is the stirring speed, r/min; T is test temperature, ℃; ω is the emulsifier dosage, ppm; t is the standing time, s; k,a,b,c,d,m and n are undetermined coefficients.

Due to the composition difference, the coefficients of different oils are different. By regressing the experiment data, these coefficients can be obtained. For the oils used in this work, the coefficients are given in Table 4.

Table 4. Model Coefficients Regression

Oil Sample No	*k*	*a*	*b*	*c*	*d*	*m*	*n*
1	45.1347	-2.9265	314.5559	-0.0006	-0.000019	0.0011	0.7931
2	71.4199	-3.0613	274.9491	-0.0004	-0.000031	0.0010	0.8514

4.2 Model Verification

For comparison, another 133 sets of experiments were conducted. Most model values fall within the +/- 20% and +/- 30% error bounds. The average and maximum relative deviations are 12.72% and 23.72% respectively.

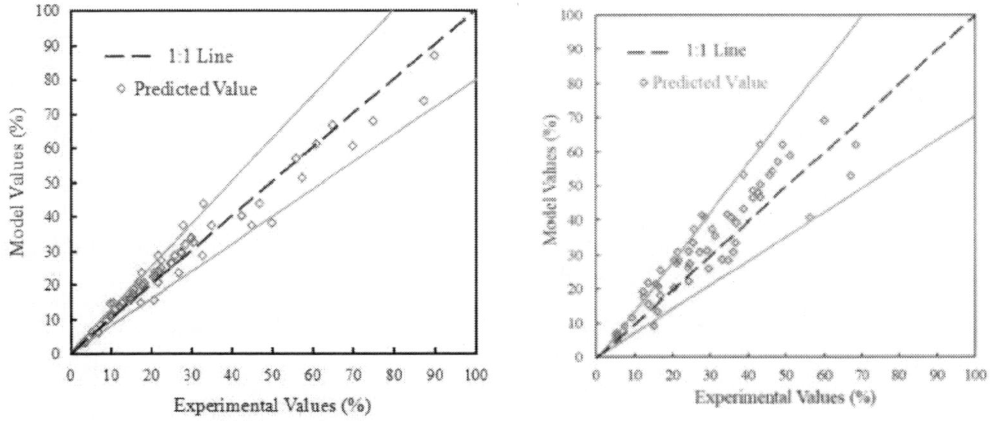

(a) (b)

Figure 11. Contrast Figure of Model Values and Experimental Values

5 Conclusions

A self-designed stirring apparatus was employed to investigate the variations of dynamic water fraction of crude oil-water systems against initial water fraction, temperature, salinity, stirring speed and emulsifier dosage. In this study, the free water fraction under flowing condition, named as the dynamic water fraction, is calculated by the equation above with t=0. It was found that the dynamic water fraction of heavy crude decreases with salinity, and emulsifier dosage, and increases with settling temperature.

A model for predicting the dynamic water fraction is accordingly established. Verification experiments show a good agreement between experimental and model values. The maximum and average relative deviations are 12.72% and 23.72% respectively.

References

1. Grassi B, Strazza D, Poesio P. Experimental validation of theoretical models in two-phase high-viscosity ratio liquid–liquid flows in horizontal and slightly inclined pipes[J]. International Journal of Multiphase Flow, 2008, 34(10): 950-965.
2. Wang W, Gong J, Angeli P. Investigation on heavy crude-water two phase flow and related flow characteristics[J]. International Journal of multiphase flow, 2011, 37(9): 1156-1164.
3. Yusuf N, Al-Wahaibi Y, Al-Wahaibi T, et al. Effect of oil viscosity on the flow structure and pressure gradient in horizontal oil-water flow[J]. Chemical Engineering Research and Design, 2012, 90(8): 1019-1030.
4. Abdurahman N H, Rosli Y M, Azhari N H, et al. Pipeline transportation of viscous crudes as concentrated oil-in-water emulsions[J]. Journal of Petroleum Science and Engineering, 2012, 90: 139-144.

5. Binner E R, Robinson J P, Silvester S A, et al. Investigation into the mechanisms by which microwave heating enhances separation of water-in-oil emulsions[J]. Fuel, 2014, 116: 516-521.
6. Binks B P, Rocher A. Effects of temperature on water-in-oil emulsions stabilised solely by wax microparticles[J]. Journal of Colloid and Interface Science, 2009, 335(1): 94-104.
7. Kang W, Guo L, Fan H, et al. Flocculation, coalescence and migration of dispersed phase
droplets and oil-water separation in heavy oil emulsion[J]. Journal of Petroleum Science and
Engineering, 2012, 81: 177-181.
8. Martínez-Palou R, Cerón-Camacho R, Chávez B, et al. Demulsification of heavy crude
oil-in-water emulsions: A comparative study between microwave and thermal heating[J]. Fuel,
2013, 113: 407-414.
9. Keleşoğlu S, Pettersen B H, Sjöblom J. Flow properties of water-in-North Sea heavy crude oil emulsions[J]. Journal of Petroleum Science and Engineering, 2012, 100: 14-23.

THERMAL DISTRIBUTION ON A HEATING FLAT PLATED COOLED WITH A SWIRLING IMPINGING JET

S. Eiamsa-ard, Mahanakorn University of Technology, Bangkok, Thailand
K. Kunnarak, Mahanakorn University of Technology, Bangkok, Thailand
K. Wongcharee, Mahanakorn University of Technology, Bangkok, Thailand
V. Chuwattanakul, King Mongkut's Institute of Technology Ladkrabang, Bangkok, Thailand

Abstract

Jet impingement cooling has been applied in various thermal engineering applications because of the efficient removal of locally concentrated heat and easy adjustment of the location where cooling is required. The present work aims to improve heat transfer uniformity by using guide-vanes to induce swirling flow to the jets. The temperature and heat transfer distributions on the impinged plate was visualized using thermochromic liquid crystal (TLC) together with image processing method. Experiments were carried out at different vane numbers of guide-vanes (N) of 4, 5, 6, 7, 8, 9 and 10 and Reynolds numbers (Re_j) of 5000, 10,000, 15,000 and 20,000 (based on the cylinder diameter; D) while the nozzle-to-plate distance ratio (l/D) was kept constant at 1.0. The thermal visualization results show that the heat transfer enhancement of the swirling impinging jet depends on the number of guide-vane (N) and jet Reynolds number (Re_j). The highest impingement heat transfer and the most uniform heat transfer is obtained by using the guide-vanes with N of 10 at Re_j of 20,000. The effect of the number of guide-vanes (N) on the uniformity of thermal distribution is also reported.

Introduction

Heat transfer enhancement technology offers numerous benefits to heat exchanger applications [1-4]. Impinging jet is one of the high potential methods for improving the heat transfer in high flux cooling or heating applications due to their high heat and mass transfer abilities. Most of typical imping jets yield extremely high heat transfer at stagnation points but heat transfer abruptly decreases in radial direction. To cope with the problem, swirling impinging jets (*SIJs*) were developed to improve the uniformity of heat transfer on impinged plates [5-10]. Wang and Mujumdar [11] studied the flow and mixing behaviors of multiple opposing jet injection and multi-set opposing jet injections with/without offset in a novel in-line mixer. Ianiro and Cardone [12] investigated the effect of swirling air jet impinging on the wall heat transfer distribution on a flat plate at different swirl numbers and nozzle-to-plate distances. The behavior of circular jets, multichannel jets, weak swirl jets and strong swirl jets were described. Li *et al.*, [13] reported the heat transfer performance of the impinging jets in a cross flow at different jet-to-cross-flow mass ratios and jet diameter ratios. The heat transfer enhancement increased with increasing jet-to-cross-flow mass ratio but decreased with increasing jet diameter. Kristiawan *et al.*, [14] used the electrodiffusion technique and the time-resolved tomographic PIV in impinging jet issued from a convergent conical nozzle to study the coherent flow structures, wall shear rate and the mass transfer phenomenon on the impinged wall. Their results revealed that the wall shear rate and local mass transfer exhibit their highest values at the normalised radial positions (r/d) of 0.6-0.7. Wongcharee *et al.*, [15, 16] investigated the effect of swirling impinging jets on the heat transfer enhancement by using nanofluids (TiO_2/CuO-water or) as the testing fluid. Impinging jets were injected at three different ratios of jet-to-plate spacing to nozzle diameter and Reynolds numbers.

The present work proposes guide-vanes as swirling flow inducers. Experiments were carried out at different vane numbers of guide-vanes (N) of 4, 5, 6, 7, 8, 9 and 10 and Reynolds numbers (Re_j) of 5000, 10,000, 15,000 and 20,000 (based on the cylinder diameter; D) while the nozzle-to-plate distance ratio (l/D) was kept constant at 1.0. A thermochromic liquid crystal (TLC) sheet was used for visualizing the temperature-field and heat transfer characteristics of swirling impinging jets (*SIJs*).

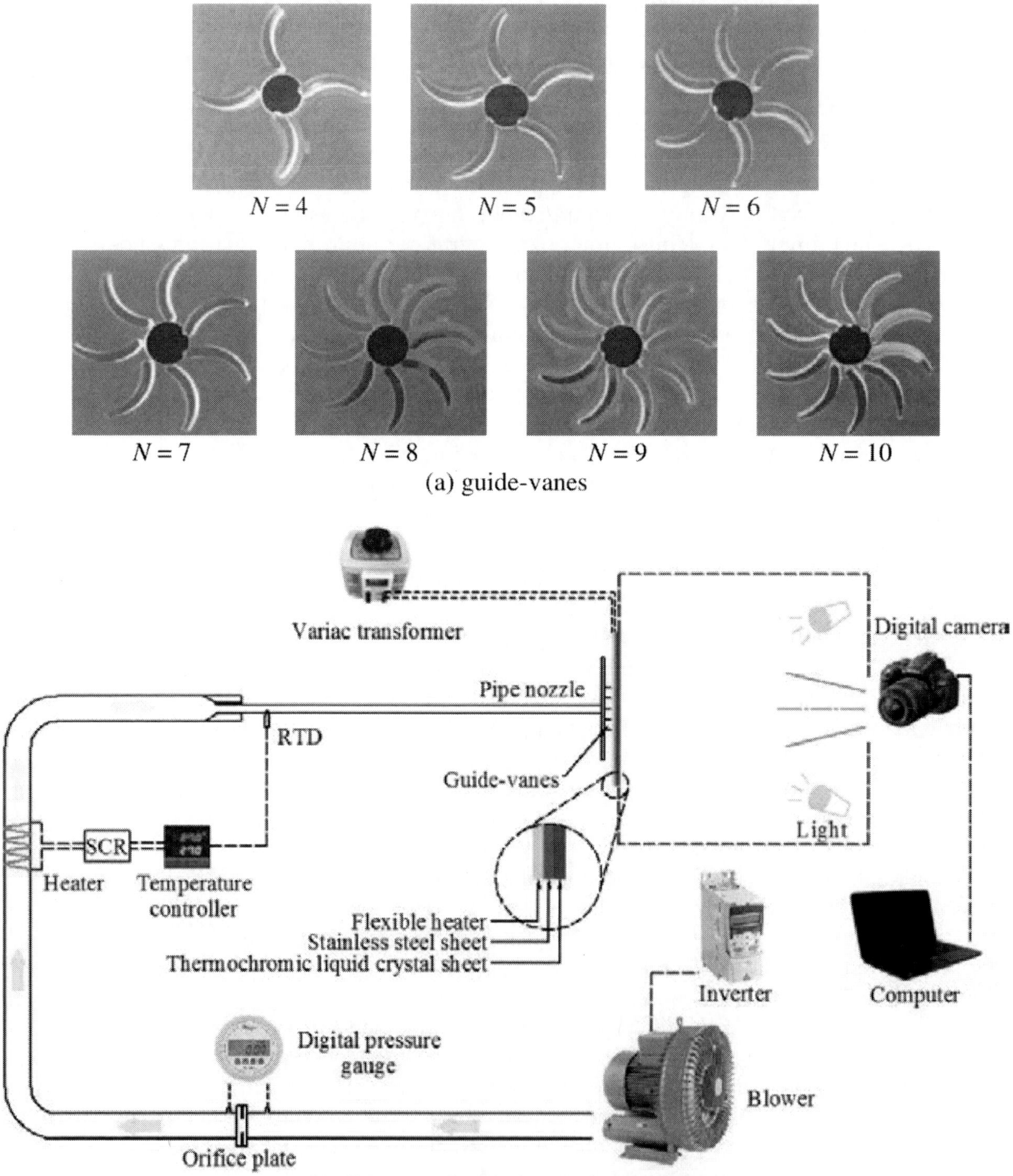

(a) guide-vanes

(b) diagram of experimental apparatus

Fig. 1. Schematic diagram of experimental apparatus.

Experimental details

The schematic of experimental setup is shown in Fig. 1. In the experiments, air at room temperature of 27°C was drawn by the 1.5 kW blower and directed to the jet chamber. Subsequently, air was heated by an electrical heater wire which was controlled by SCR (Silicon-Controlled Rectifier) via a temperature controller. The volumetric flow rate of the air through the pipe nozzle was controlled corresponding to the desired Reynolds number. The pipe nozzle, in which the guide-vanes located had an inner diameter of 14 mm and a length of 1000 mm. The guide-vane was made of an acrylic with thickness of 2 mm and 82 mm in vane diameter. The outer pipe nozzle surface and the jet chamber were well insulated to minimize heat loss to surroundings. The pressure drop of air across the orifice plate was measured by using a U-tube manometer. The air delivered from the nozzle pipe was then normally impinged on the impinged plate which was made of a stainless-steel foil (SUS304) with length, width and thickness of 250 mm, 250 mm and 0.03 mm, respectively. A thermochromic liquid crystal (TLC) sheet was attached on the rear of the stainless foil with binder. A TLC sheet is temperature sensitive, its color changes from black, brown, red, yellow, green to blue from 29°C to 36°C.

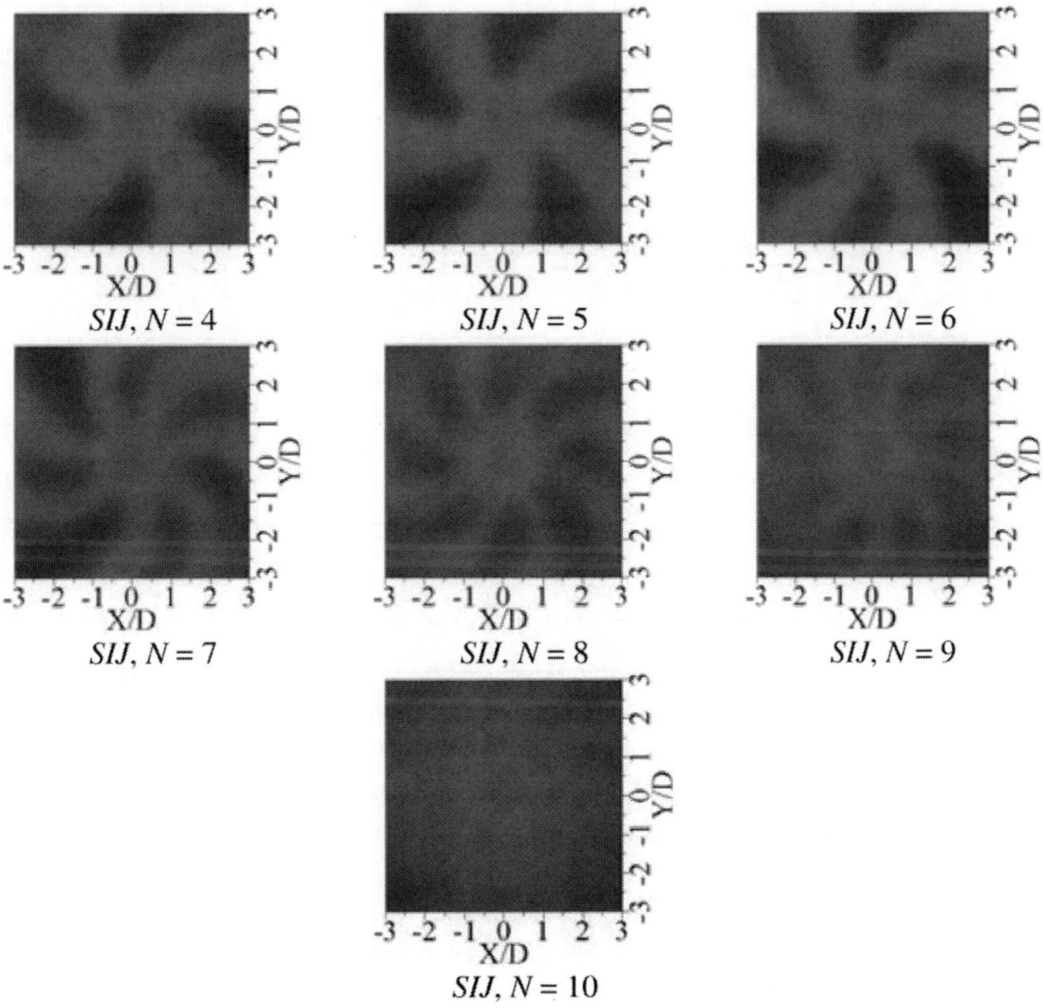

Fig. 2. Effect of vane number (*N*) on thermal distribution of a heating flat plated cooled by impinging jets.

Direct current was supplied to the stainless-steel foil from a power supply unit via the copper bus bar electrodes, which were attached to a stainless-steel foil and an acrylic plate. The input power to the stainless-steel foil was evaluated from the power supply electric current and resistance of the stainless foil. The experimental results indicated the impinged surface was maintained under a uniform heat flux condition. When the temperature distribution on impinged surface, as indicated by the color pattern of the TLC, reached a steady state, the color pattern on thermochromic liquid crystal (TLC) sheet was recorded by a digital camera. Because the stainless-steel foil used in this study was extremely thin, the temperature of the target could be assessed from the color pattern of the TLC sheet on the rear side. The recorded image was then converted from RGB (red, green, blue) to HSI (hue, saturation, intensity) color domain. Subsequently, the distribution of hue was converted to surface temperature via calibration curve between hue and temperature. The correlation of the curve was achieved by calibrating thermocouples. Heat transfer results are presented in the form of Nusselt number which could be calculated from a convective heat transfer coefficient. The local convective heat transfer coefficient and local Nusselt number are determined from:

$$h = q/T_w - T_j \qquad (1)$$

$$Nu = hD/k \qquad (2)$$

where q is the heat flux supplied to impinged plate. T_j is the jet temperature measured in settling chamber, T_w is the local temperature on the impinged plate measured by the TLC sheet.

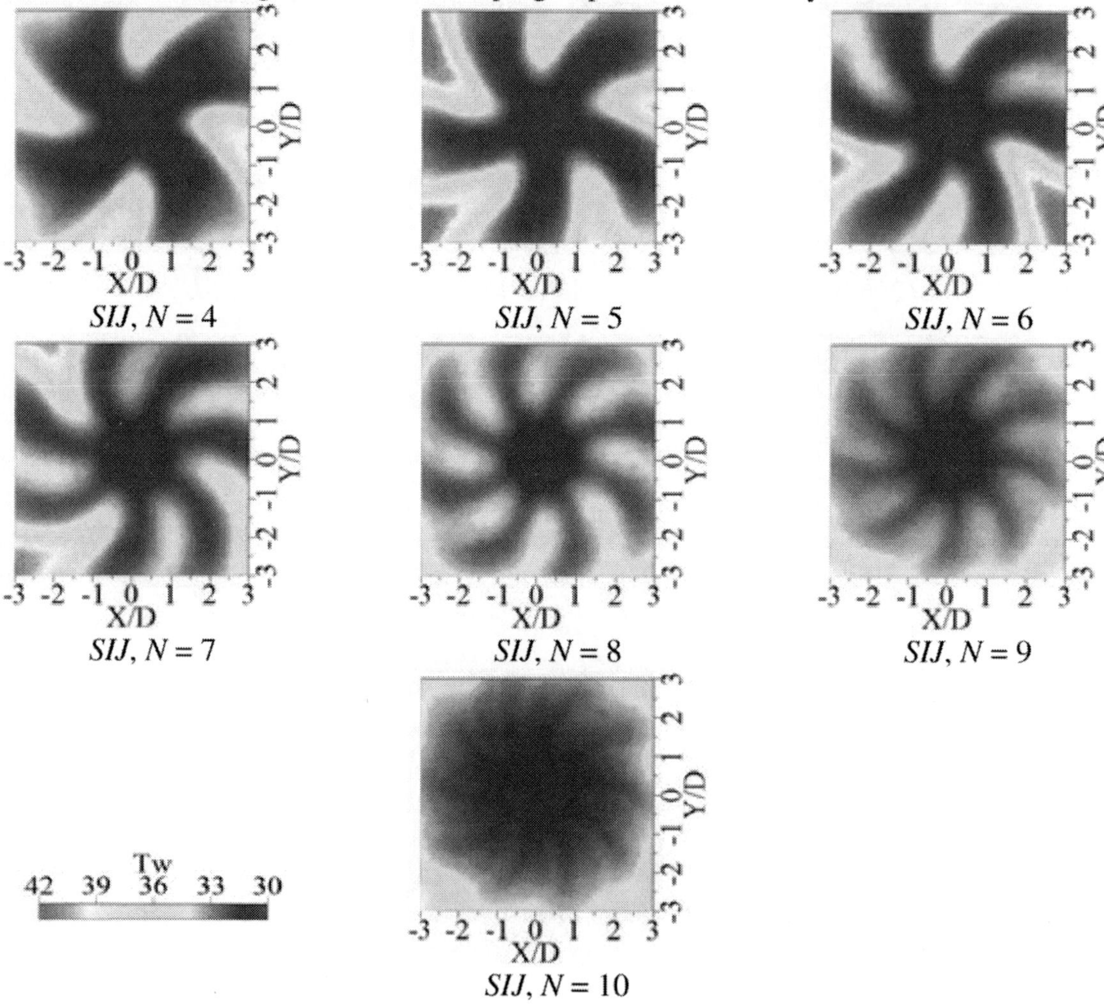

Fig. 3. Effect of vane number (*N*) on temperature distribution of a heating flat plated.

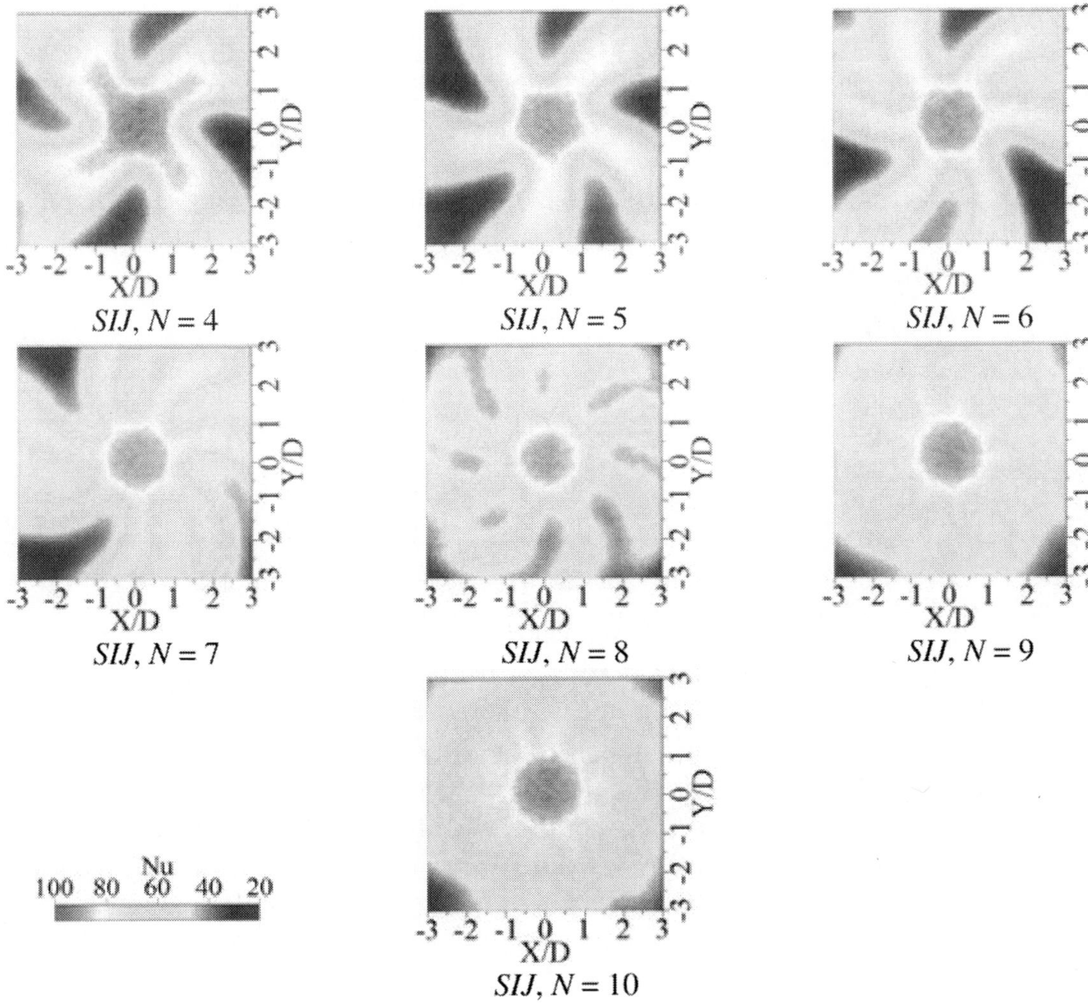

Fig. 4. Effect of vane number (*N*) on Nusselt number distribution of a heating flat plated cooled by impinging jets.

Experimental results

The temperature and Nusselt number distributions on the impinged plate of swirling impinging jets (SIJs) are demonstrated in Figs. 2-4. Obviously, the temperatures around stagnation point (X/D = 0.0 and Y/D = 0.0) are low corresponding to the high local Nusselt numbers around the point. The SIJs issuing through guide-vanes with vane numbers (*N*) of 7, 8 and 9 have low uniformities of temperature and Nusselt number distributions which are even lower than that of the higher vane numbers (*N*) because of the large gaps between the adjacent vanes. On the other hand, The *SIJs* issuing through guide-vanes with vane numbers (*N*) of 4 and 10 possess better uniformities of temperature and Nusselt number distributions than the guide-vanes with vane numbers (*N*) of 5, 6, 7, 8 and 9. The better uniformities at more vane numbers are attributed to the more consistent swirling intensities. The highest impingement heat transfer and the most uniform radial heat transfer is obtained by using the guide-vanes with *N* of 4 and 10 at jet Reynolds number of 20,000. Therefore, it can be concluded that the use swirling flow inducer with proper geometry in jet impingement improves the uniformity of radial heat transfer distribution.

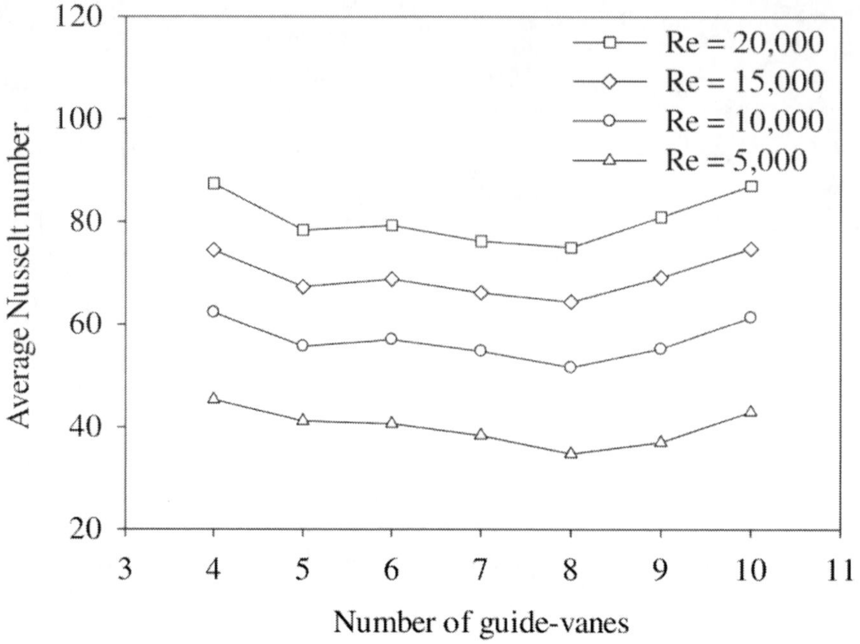

Fig. 5. Effect of vane number (*N*) on average Nusselt number at r/D= 2.0.

Results the influence of the vane numbers of guide-vanes ($N = 4, 5, 6, 7, 8, 9$ and 10) on Nusselt number (*Nu*) are showed in Fig. 5. Eventually, as vane numbers increased, Nusselt number (*Nu*) increased for all cases especially for the *SIJs*. For the swirling impinging jets (SIJs), the average Nusselt number (*Nu*) provided by vane number of $N = 4$ and 10 are 12.7% and 10.7% higher than that the lowest vane number of $N = 9$. However, at vane numbers of $N = 5, 6, 7, 8$ and 9 yield lower average Nusselt number (*Nu*) than the corresponding the highest vane number of $N = 4$ by around 9.14%,7.91%, 12.7%, 18.5% and 10.7%, respectively.

Conclusions

An experimental investigation has been performed to study the heat transfer rate in the impingement plate by means of guide-vanes (swirling impinging jets, SIJs) mounted at the exit of a nozzle pipe. In the present work, the swirling impinging jets (*SIJs*) at different vane numbers of guide-vanes ($N = 4, 5, 6, 7, 8, 9$ and 10) were carried-out at Reynolds numbers (Re_j) of 5000, 10,000, 15,000 and 20,000 (based on the cylinder diameter; *D*). It is found that the mean Nusselt number of $Re_j = 20{,}000$ may be increased by 103%, 74% and 42.3% over than that $Re_j = 5000$, 10,000 and 15,000 at vane number of $N = 10$. The increase in heat transfer can be explained by the swirling impinging jets (SIJs) as a result of the high momentum flows of the fluid and induce flow around the impingement plate to hit.

References

1. Xu, L., Lan, J., Ma, Y., Gao, J. and Li, Y. (2017), "Numerical Study on Heat Transfer by Swirling Impinging Jets issuing from a Screw-Thread Nozzle," International Journal of Heat and Mass Transfer, December, pp. 232-237.
2. Nuntadusit, C., Wae-hayee, M., Bunyajitradulya, A. and Eiamsa-ard, S. (2012), "Heat Transfer

Enhancement by Multiple Swirling Impinging Jets with Twisted-Tape Swirl Generators," International Communications in Heat and Mass Transfer, January, pp. 102-107.

3. Ahmed, Z.U., Al-Abdeli, Y.M. and Guzzomi, F.G. (2016), "Heat Transfer Characteristics of Swirling and Non-Swirling Impinging Turbulent Jets," International Journal of Heat and Mass Transfer, November, pp. 991-1003.
4. Singh, P. and Chander, S. (2018), "Heat Transfer and Fluid Flow Characteristics of a Pair of Interacting Dual Swirling Flame Jets Impinging on a Flat Surface," International Journal of Heat and Mass Transfer, September, pp. 90-108.
5. Eiamsa-ard, S., Nanan, K. and Wongcharee, K. (2015), "Heat Transfer Visualization of Co/Counter-Dual Swirling Impinging Jets by Thermochromic Liquid Crystal Method," International Journal of Heat and Mass Transfer, July, pp. 600-621.
6. Hindasageri, V., Vedula, R.P. and Prabhu, S.V. (2015), "Heat Transfer Distribution of Swirling Flame Jet Impinging on a Flat Plate using Twisted Tapes," International Journal of Heat and Mass Transfer, December, pp. 1128-1139.
7. Nuntadusit, C., Wae-hayee, M., Bunyajitradulya, A. and Eiamsa-ard, S. (2012), "Visualization of Flow and Heat Transfer Characteristics for Swirling Impinging Jet," International Communications in Heat and Mass Transfer, May, pp. 640-648.
8. Wannassi, M. and Monnoyer, F. (2015), "Fluid Flow and Convective Heat Transfer of Combined Swirling and Straight Impinging Jet Arrays," Applied Thermal Engineering, March, pp. 62-73.
9. Nanan, K., Wongcharee, K., Nuntadusit, C. and Eiamsa-ard, S. (2012), "Forced Convective Heat Transfer by Swirling Impinging Jets issuing from Nozzles Equipped with Twisted Tapes," International Communications in Heat and Mass Transfer, July, pp. 844-852.
10. Nanan K. and Eiamsa-ard, P. (2014), "Heat Transfer of Swirling Jet Impinging on a Flat Surface with Swirl Generators," Transactions of the TSME, Journal of Research and Applications in Mechanical Engineering, pp. 103-110.
11. Wang, S.J. and Mujumdar, A.S. (2007), "Flow and Mixing Characteristics of Multiple and Multi-Set Opposing Jets," Chemical Engineering and Processing: Process Intensification, August, pp. 703-712.
12. Ianiro, A. and Cardone, G. (2012), "Heat Transfer Rate and Uniformity in Multichannel Swirling Impinging Jets," Applied Thermal Engineering, December, pp. 89-98
13. Li, G., Zheng, Y., Hu, G. and Zhang, Z. (2014), "Convective Heat Transfer Enhancement of a Rectangular Flat Plate by an Impinging Jet in Cross Flow," Chinese Journal of Chemical Engineering, May, pp. 489-495.
14. Kristiawan, M., Sodjavi, K., Montagné, B., Meslem, A. and Sobolik, V. (2015), "Mass Transfer and Shear Rate on a Wall Normal to an Impinging Circular Jet," Chemical Engineering Science, August, pp. 32-45.
15. Wongcharee, K., Chuwattanakul, V. and Eiamsa-ard, S. (2017), "Heat Transfer of Swirling Impinging Jets with TiO_2-Water Nanofluids," Chemical Engineering and Processing: Process Intensification, April, pp. 16-23.
16. Wongcharee, K., Chuwattanakul, V. and Eiamsa-ard, S. (2017), "Influence of CuO/Water Nanofluid Concentration and Swirling Flow on Jet Impingement Cooling," International Communications in Heat and Mass Transfer, November, pp. 277-283.

PERIODICALLY FULLY DEVELOPED LAMINAR FLOW AND HEAT TRANSFER CHARACTERISTICS IN A TUBE INSERTED WITH RECTANGULAR-CUT TWISTED TAPE

V. Kongkaitpaiboon, Mahanakorn University of Technology, Bangkok, Thailand
A. Saysroy, Phetchaburi Rajabhat University, Phetchaburi, Thailand
K. Wongcharee, Mahanakorn University of Technology, Bangkok, Thailand
V. Chuwattanakul, King Mongkut's Institute of Technology Ladkrabang, Bangkok, Thailand
M. Pimsarn, King Mongkut's Institute of Technology Ladkrabang, Bangkok, Thailand
S. Eiamsa-ard, Mahanakorn University of Technology, Bangkok, Thailand

Abstract

Periodically fully developed three-dimensional flow and heat transfer in the tubes inserted with rectangular-cut twisted tapes are numerically studied. The investigation covers the tapes with perforated width to tape width ratios ($WR = w/W = 0.7$, 0.8 and 0.9) and perforated length to tape width ratios ($LR = L/W = 0.5$, 0.6, 0.7, 0.8 and 0.9). Simulation is performed in laminar regime at Reynolds number varying from 400 to 2000. The thermos-hydraulic performances of the tubes inserted with rectangular-cut twisted tapes are compared with those of the plain tube in terms of Nusselt number (Nu/Nu_p), friction factor (f/f_p), and thermal performance factor. At a given Reynolds number, heat transfer, friction loss and thermal enhancement factor (*TEF*) increase with decreasing perforated width to tape width ratio (*WR*) perforated length to tape width ratio (*LR*). The tube with tape at the smallest *WR* of 0.7 and the smallest *LR* of 0.5 offers the maximum heat transfer enhancement which is as high as 17.2 times of the plain tube which corresponds to the maximum thermal performance factor of 8.66.

Introduction

Inserting twisted tape is one of the most effective passive heat transfer enhancement methods. Twisted tapes function as swirl generators which help in interrupting the thermal/hydrodynamic boundary layers and facilitate heat transfer through the layers. Twisted tapes are implemented with the expense of addition energy consumption due to increase friction loss. Therefore, developing twisted tape for the reasonable tradeoff between the enhanced heat transfer and increased friction loss is necessary for practical use. In common, typical twisted tapes were modified by introducing spaces on their blades to decrease the friction loss of fluid flow. He *et al.* [1] used cross hollow twisted tapes with hollow widths of 6, 8 and 10 mm under uniform heat flux. Their results revealed that the Nusselt number (*Nu*) decreased as the hollow width increased. The performance evaluation criterion (*PEC*) varied from 0.87 to 0.98 for Reynolds numbers of 5600-18,000. Thianpong *et al.* [2] proposed perforated twisted-tapes with parallel wings. The geometric parameters examined were hole diameter ratios (d/W) of 0.11, 0.33 and 0.55 and wing depth ratios (w/W) of 0.11, 0.22 and 0.33. The tubes with perforated twisted-tapes with parallel wings yielded heat transfer enhancement up to 208% as compare to the plain tube. The tape with $d/W = 0.11$ and $w/W = 0.33$, gave the maximum thermal performance factor of 1.32, at Reynolds number of 5500. Thianpong *et al.* [3] also studied the heat transfer enhancement of perforated twisted tapes with $y/W = 3.0$, 4.0 and 5.0, $d/W = 0.11$, 0.14 and 0.17 and $s/W = 0.4$, 0.6 and 0.8 where y is a twist length, d is a perforation hole diameter, s is a spacing between holes (pitch) and W is a tape width. The experimental results revealed that Nusselt number increased with decreasing s/W and y/W and increasing d/W. The maximum heat transfer rate was obtained by utilizing the tape with $s/W = 0.4$, $d/W = 0.17$ and *y/W* = 3.0, which is higher than those obtained from the plain tube by 86.7%. Nanan *et al.* [4] investigated the effect of diameter ratio and pitch ratio of perforated helical twisted-tape inserts on heat transfer enhancement. The experimental results revealed that Heat transfer, friction loss and thermal performance factor increased as diameter ratio decreased and pitch ratio increased. It was also found that the use of perforated helical twisted-tape inserts led to the reduction of friction loss as compare to that of the typical helical twisted tape. Eiamsa-ard and Promvonge [5]proposed double-sided delta-wing tape insert with alternate-axes transfer for heat transfer enhancement in heat exchanger tube. It was found that at the optimum condition the tube with the modified tape gave the maximum thermal performance factor of 1.19. Hasanpour *et al.* [6] examined the heat transfer rate and pressure loss in a corrugated tube inserted with perforated/V-cut/U-cut twisted tapes with the following geometric parameters: twist ratios of 3.0, 5.0 and 7.0, hole diameter ratios of 0.11 and 0.33, width and depth ratios of the cuts of 0.3 to 0.6. They found that V-cut/U-cut twisted tapes gave better heat transfer rate and caused higher pressure loss than the typical twisted tape while perforated twisted tape gave the opposite results

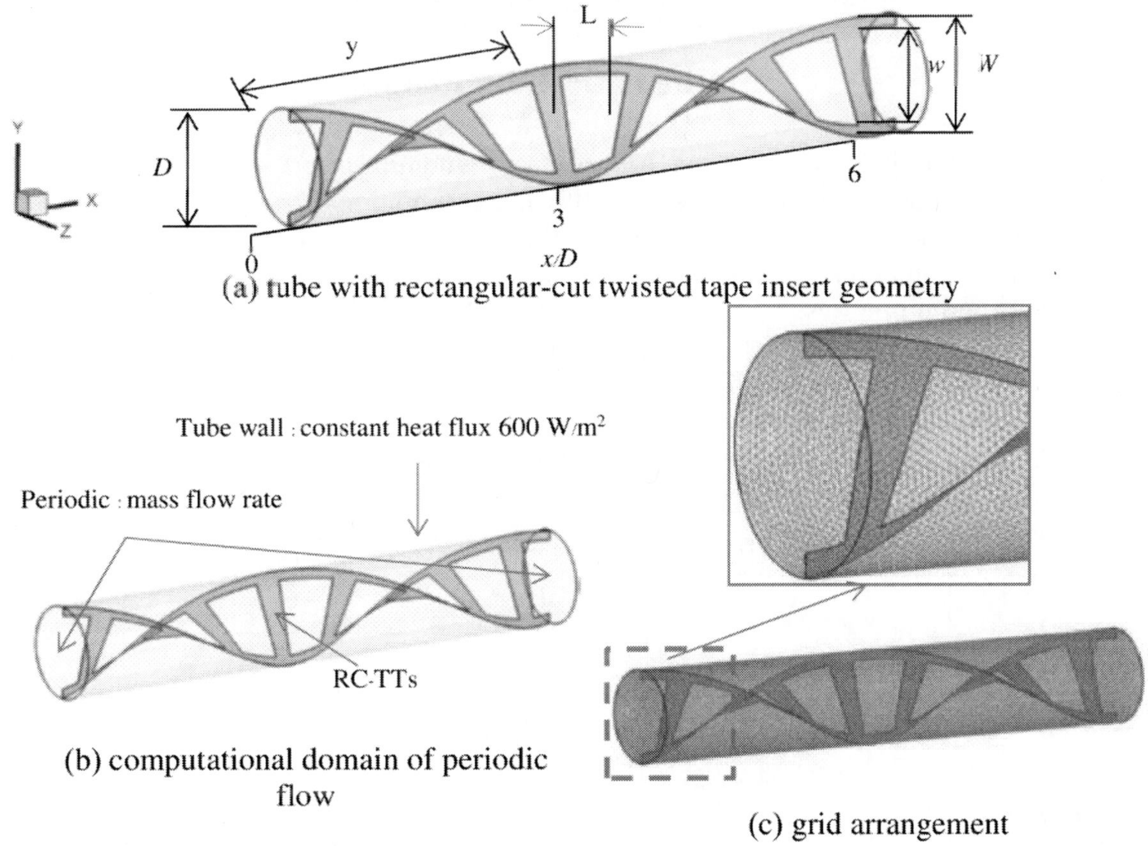

(a) tube with rectangular-cut twisted tape insert geometry

(b) computational domain of periodic flow

(c) grid arrangement

Figure 1. Tube inserted with rectangular-cut twisted tape, computational domain of periodic flow, and grid arrangement.

Above literature review shows that the performance of twisted tapes is strongly dependent on tape geometries and most of modified twisted tapes perform better that typical twisted tape. However, most of the researches relating the use of twisted tapes for heat transfer enhancement are experimental studies. The reports of heat transfer phenomena are limited. To gain a deeper insight of flow field, fluid temperature, local Nusselt number, pressure loss and thermal performance characteristics, the present work proposes a numerical simulation of the behaviors in circular tubes inserted with rectangular-cut twisted tapes. The rectangular-cuts are introduced on the twisted tapes in order to reduce pressure loss of the fluid flow. Fifteen tapes with different perforated width to tape width ratios ($WR = w/W$) and perforated length to tape width ratios ($LR = L/W$) are examined to find the optimum tape geometry. The plain tube and also the tube with a typical twisted tape are also studied for comparison.

Rectangular-cut twisted tape geometry

The systems of interest is horizontal tubes mounted with rectangular-cut twisted tapes. The details of a tube inserted with rectangular-cut twisted tape swirl generators are displayed in Fig. 1. The tapes with the perforated width to tape width ratios ($WR = w/W = 0.7$, 0.8 and 0.9) and perforated length to tape width ratios ($LR = L/W = 0.5$, 0.6, 0.7, 0.8 and 0.9) of the rectangular-cut twisted tapes are examined. All rectangular-cut twisted tapes have the same tape twist ratio (y/W) of 3.0 and all tubes have the same diameter. The simulation is carried out for laminar flow at Reynolds numbers (Re) varying from 400 to 2000 using water as the working fluid. The tube wall is subjected to a constant wall heat flux condition at 600 W/m^2 while the flow is under periodic flow condition.

Mathematical foundation

The numerical model for fluid flow and heat transfer in a round tube inserted with rectangular-cut twisted tape swirl generators was developed under the following assumptions:

- The flow is steady three-dimensional (3-D) fluid flow and heat transfer.
- The flow is laminar and incompressible.
- Constant fluid properties.
- Body forces and viscous dissipation are ignored.
- Negligible radiation heat transfer.

Based on the above assumptions, the tube flow is governed by the continuity, the Navier-Stokes equations and the energy equation. In the Cartesian tensor system these equations can be written as follows:

Continuity equation:

$$\frac{\partial}{\partial x_i}(\rho u_i) = 0 \tag{1}$$

Momentum equation:

$$\frac{\partial(\rho u_i u_j)}{\partial x_j} = -\frac{\partial p}{\partial x_i} + \frac{\partial}{\partial x_j}\left[\mu\left(\frac{\partial u_i}{\partial x_j} + \frac{\partial u_j}{\partial x_i}\right)\right] \tag{2}$$

Energy equation:

$$\frac{\partial}{\partial x_i}(\rho u_i T) = \frac{\partial}{\partial x_j}\left(\Gamma \frac{\partial T}{\partial x_j}\right) \tag{3}$$

where Γ is the thermal diffusivity which is defined as

$$\Gamma = \frac{\mu}{Pr} \tag{4}$$

The governing equations were discretized by the QUICK differencing scheme, decoupling with the SIMPLE algorithm and solved using a finite volume approach (Patankar [7]). The solutions were considered to be converged when the normalized residual values were less than 10^{-6} for all variables.

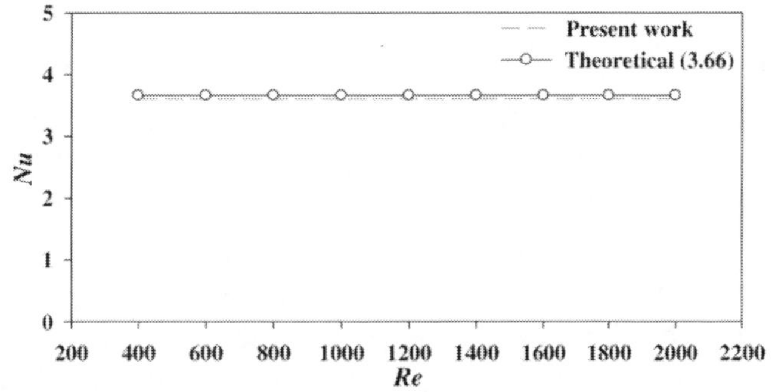

(a) *Nu*

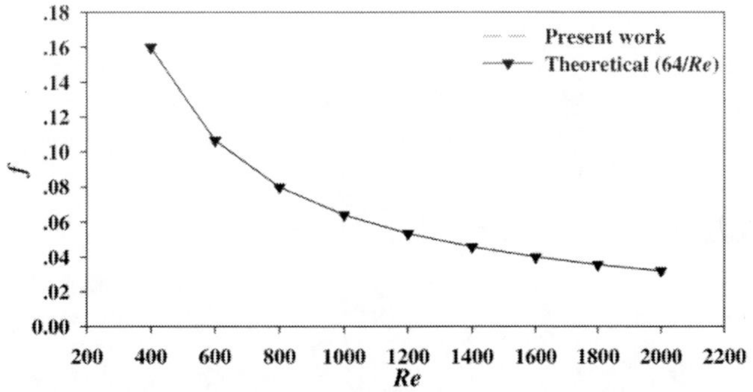

(b) *f*

Figure 2. Variations of Nusselt number and friction factor with Reynolds number for the plain tube.

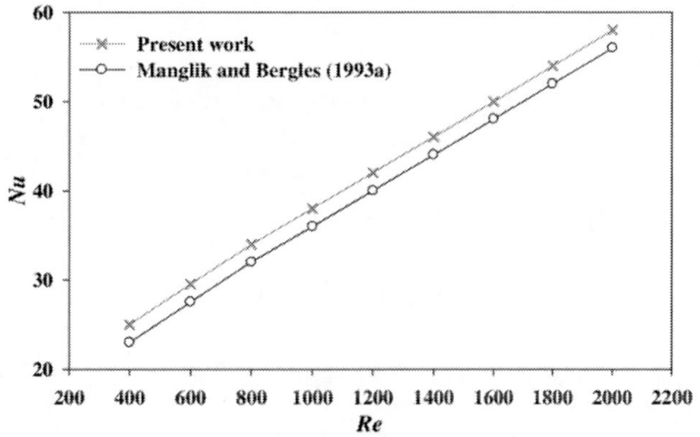

(a) *Nu*

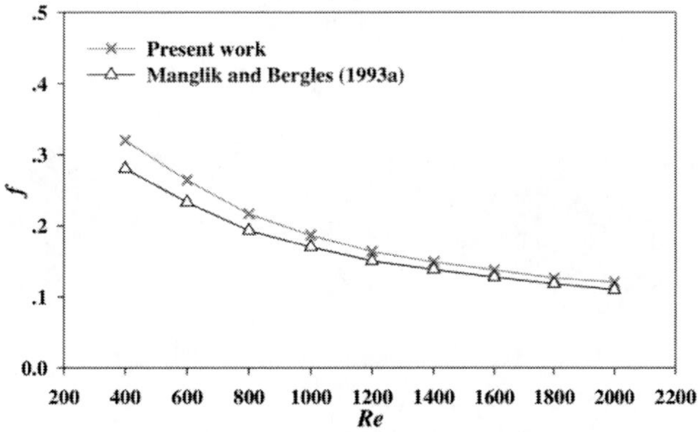

(b) *f*

Figure 3. Variations of Nusselt number and friction factor with Reynolds number for the tube with typical twisted tape insert.

Numerical results

Verification of numerical method

Prior to the simulation for the tubes inserted with rectangular-cut twisted-tape, the numerical results were validated by comparing the present results (Nusselt number and friction factor) of the plain tube with those of the previous standard correlations which are available in the open literature by Incropera *et al.* [8]. Figure 2 shows that the deviations of the present numerical Nusselt number (Nu) and friction factor (f) results for deviate from previous standard correlations within ±1.53% and ±0.1%, respectively. The present numerical results of the plain tube inserted with the typical twisted tape were validated with the standard correlations of Manglik and Bergles [9] under similar conditions in laminar regions. Figure 3 shows that the deviations of the present Nu and f from those of Manglik and Bergles are within 6.19% and 13.3%, respectively.

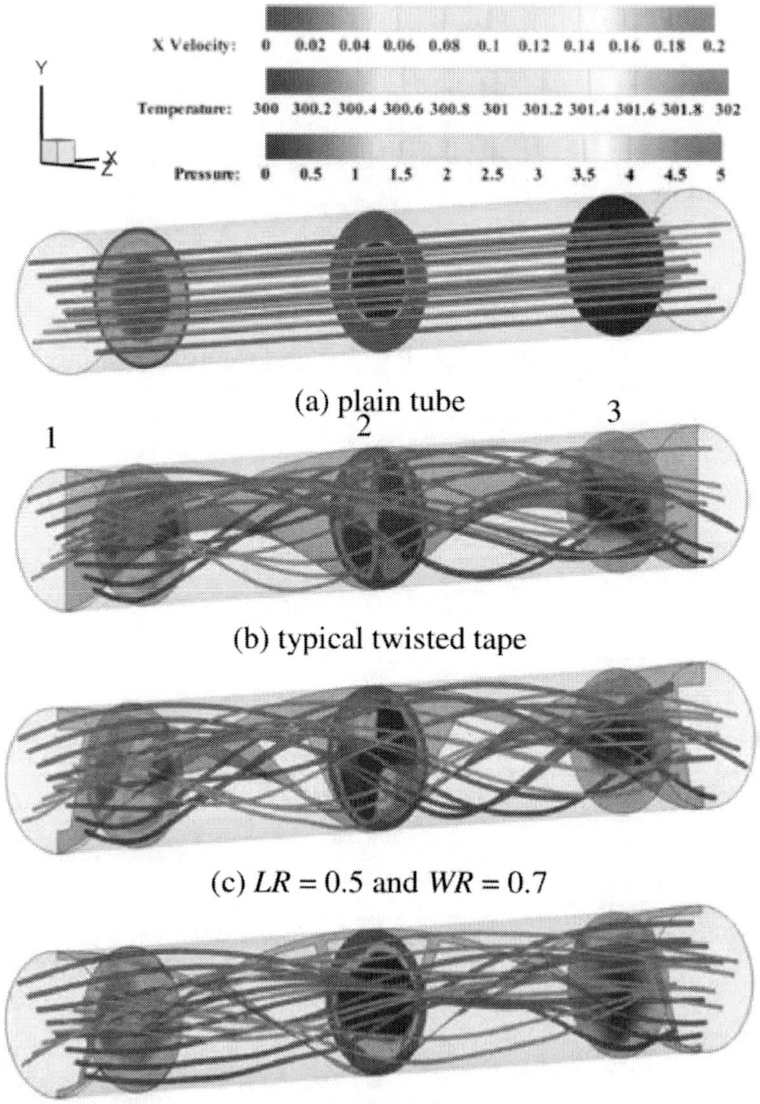

(a) plain tube

(b) typical twisted tape

(c) $LR = 0.5$ and $WR = 0.7$

(d) $LR = 0.9$ and $WR = 0.9$

Figure 4. The contour plots of streamline, velocity field (plane 1), temperature field (plane 2), and pressure field (plane 3) for: (a) plain tube, (b) typical twisted tape, (c) $LR = 0.5$ and $WR = 0.7$, and (d) $LR = 0.9$ and $WR = 0.9$.

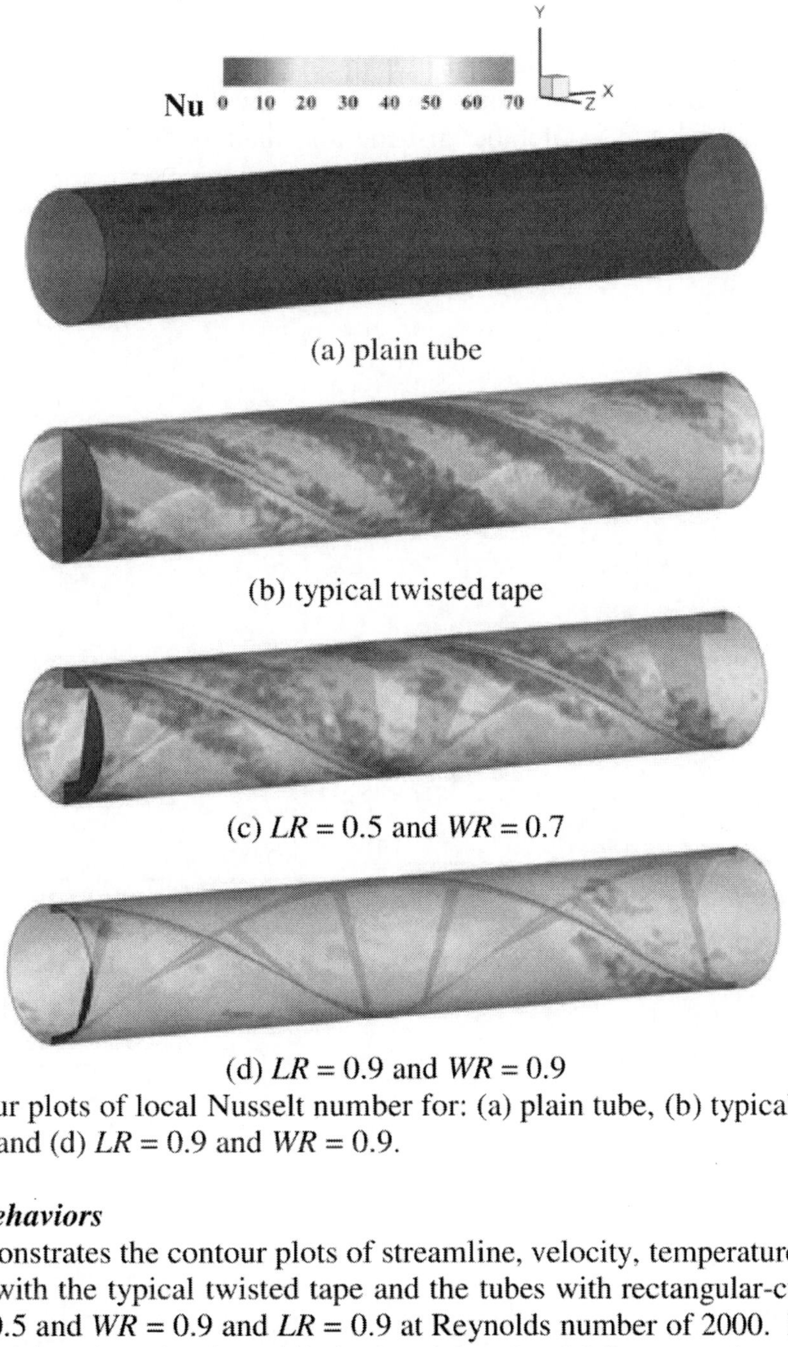

(a) plain tube

(b) typical twisted tape

(c) LR = 0.5 and WR = 0.7

(d) LR = 0.9 and WR = 0.9

Figure 5. The contour plots of local Nusselt number for: (a) plain tube, (b) typical twisted tape, (c) LR = 0.5 and WR = 0.7, and (d) LR = 0.9 and WR = 0.9.

Flow and thermal behaviors

Figure 4 demonstrates the contour plots of streamline, velocity, temperature, and pressure in the plain tube, the tube with the typical twisted tape and the tubes with rectangular-cut twisted tapes with WR = 0.7 and LR = 0.5 and WR = 0.9 and LR = 0.9 at Reynolds number of 2000. Evidently, only axial flow is found in the plain tube (Fig. 4a) while both axial and swirl flows are detected in the tubes with typical twisted tape (Fig. 4b) and rectangular-cut twisted tape inserts (Fig. 4c and d). The created swirl flows help in promoting disturbance of a thermal boundary layer and the transportation of the fluid between the core and near-wall regions. The insertion of twisted tapes leads to the promoting velocity and reducing fluid temperature near wall regions, improving fluid mixing between core and near wall regions and thus thinning thermal boundary layers as compared to those in the plain tube.

Figure 4 also shows the effect of swirl flow is more notable as perforated width to tape width ratios (WR) and perforated length to tape width ratios (LR) decrease since the tube with the typical twisted tape and the one with the rectangular-cut twisted tape with the small perforated width to tape width ratio (WR = 0.7) and perforated length to tape width ratio (LR = 0.5) yield thinner thermal boundary layers than the one with the large perforated width to tape width ratio (WR = 0.9) and perforated length to tape width ratio (LR = 0.9). These results accord with those in Fig. 5 which shows that the tube with the typical twisted tape and the one with the rectangular-cut twisted tape with the small perforated width to tape width ratios (WR = 0.7) and perforated length to tape width ratios (LR = 0.5), give higher Nusselt numbers and more uniform Nusselt number distributions than the one with WR = 0.9 and LR = 0.9.

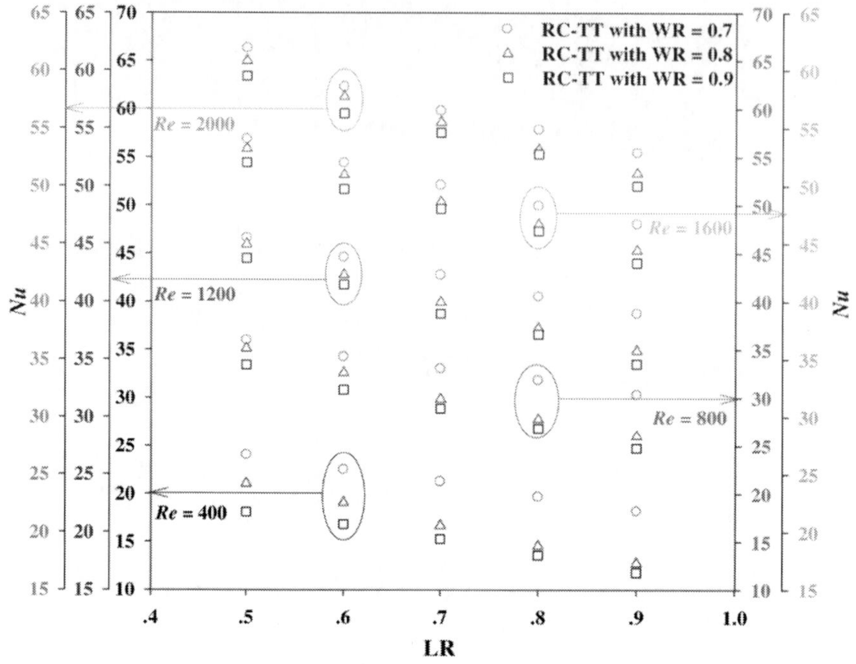

(a) Nu

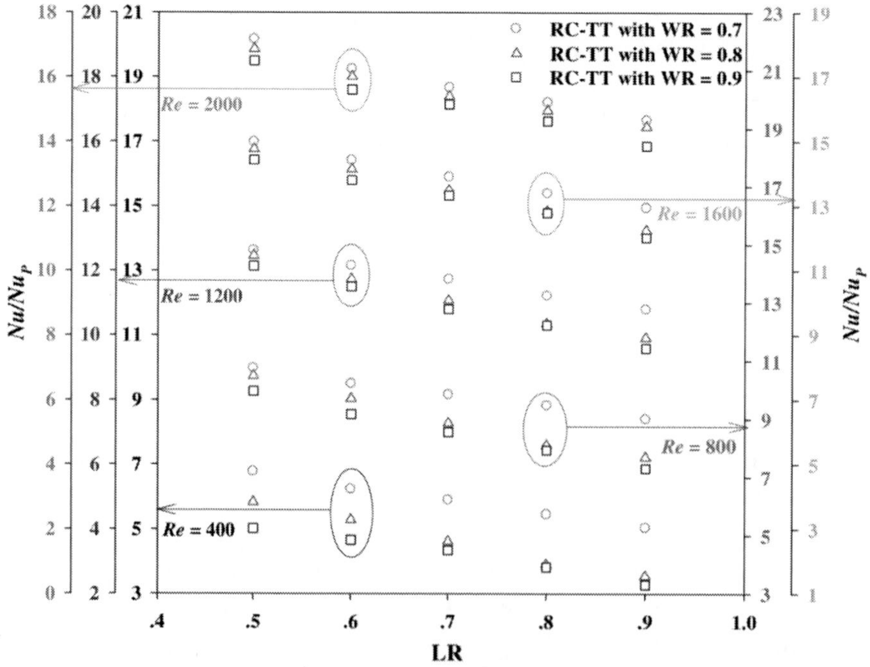

(b) Nu/Nu_p

Figure 6. Influence of perforated length to tape width ratio (LR) and perforated width to tape width ratio (WR) on heat transfer at different Reynolds numbers (Re) for (a) Nu and (b) Nu/Nu_p.

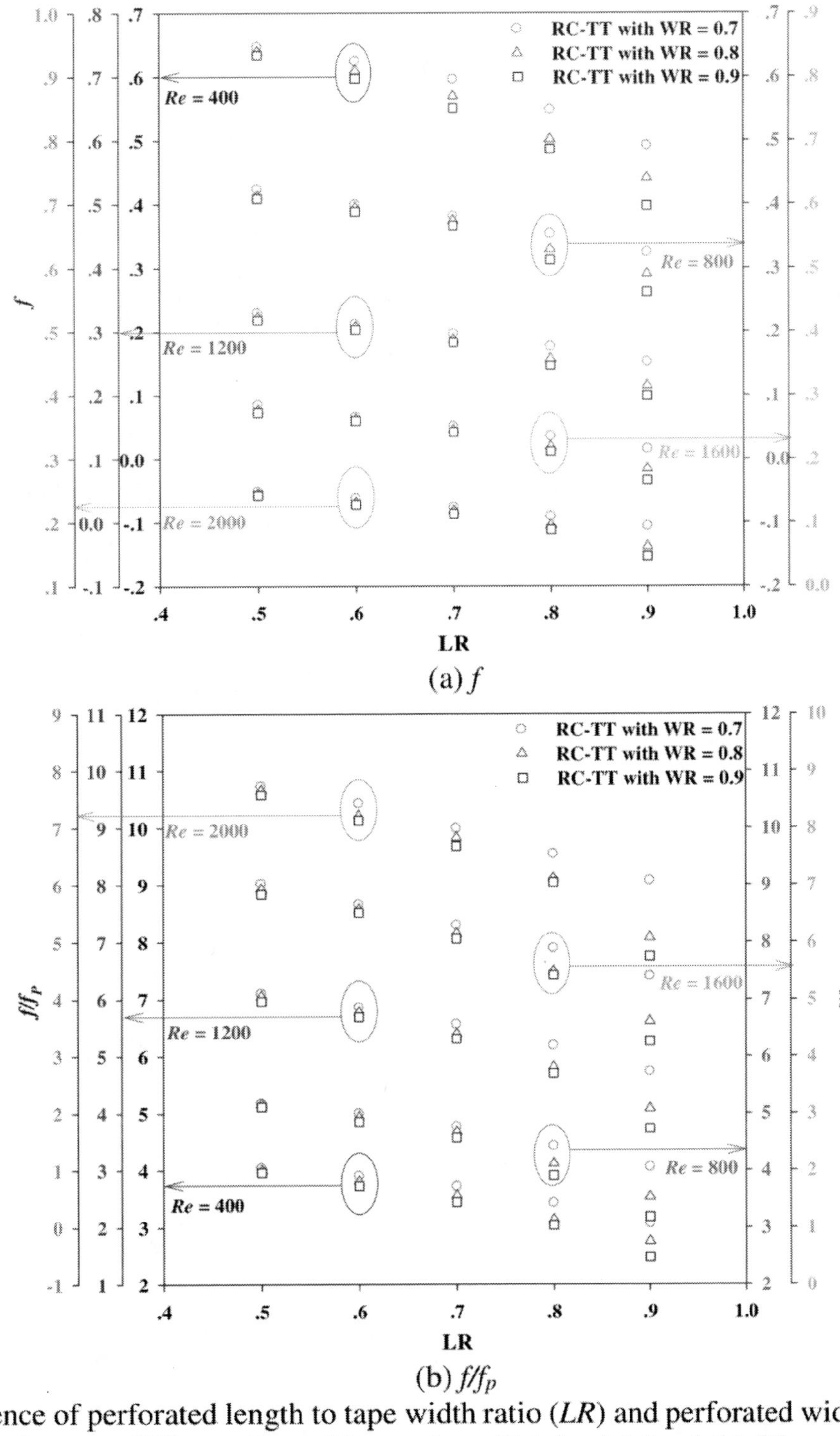

Figure 7. Influence of perforated length to tape width ratio (*LR*) and perforated width to tape width ratio (*WR*) on friction factor at different Reynolds numbers (*Re*) for (a) f and (b) f/f_p.

Heat transfer

Figure 6 presents the effect of perforated width of cut ratios (*WR* = 0.7, 0.8 and 0.9) and perforated length to tape width ratios (*LR* = 0.5, 0.6, 0.7, 0.8 and 0.9) on Nusselt number and Nusselt number ratio of the tubes with rectangular-cut twisted tape inserts. At a given Reynolds number, Nusselt number and Nusselt number ratio increase with decreasing perforated width to tape width ratio (*WR*) perforated length to tape width ratio (*LR*).This is attributed the higher turbulence intensity induced by the tape with smaller perforated width and length since the tape possesses larger surface area, thus more fluid is perturbed by the tape. According to the obtained results, the maximum Nusselt numbers of the tubes with the rectangular-cut twisted tapes with *LR* = 0.5, 0.6, 0.7, 0.8 and 0.9 are, respectively, 17.2, 16.3, 15.6, 15.2 and 14.6 times of those of the plain tube.

Pressure

The variations of friction factor (f) and friction factor ratio (f/f_p) with Reynolds number (*Re*) of the tubes with rectangular-cut twisted tapes with perforated width to tape width ratios (WR) of 0.7, 0.8 and 0.9 and perforated length to tape width ratios (*LR* = 0.5, 0.6, 0.7, 0.8 and 0.9) are shown in Fig. 7. For all cases, friction factor (f) tends to decrease while friction factor ratio (f/f_p) tends to increase with the rise of Reynolds number (*Re*). At a given Reynolds number, friction factor (f) and friction factor ratio (f/f_p) increase with decreasing perforated width to tape width ratio (*WR*) perforated length to tape width ratio (*LR*) since the tape with larger surface area causes higher turbulence intensity (in other words, the smaller perforation allows less fluid to flow freely through the perforated part). The maximum friction factors of the tubes with rectangular-cut twisted tape swirls with *LR* = 0.5, 0.6, 0.7, 0.8 and 0.9, are, respectively 7.7, 7.4, 7.0, 6.5 and 6.1 times of those of the plain tube.

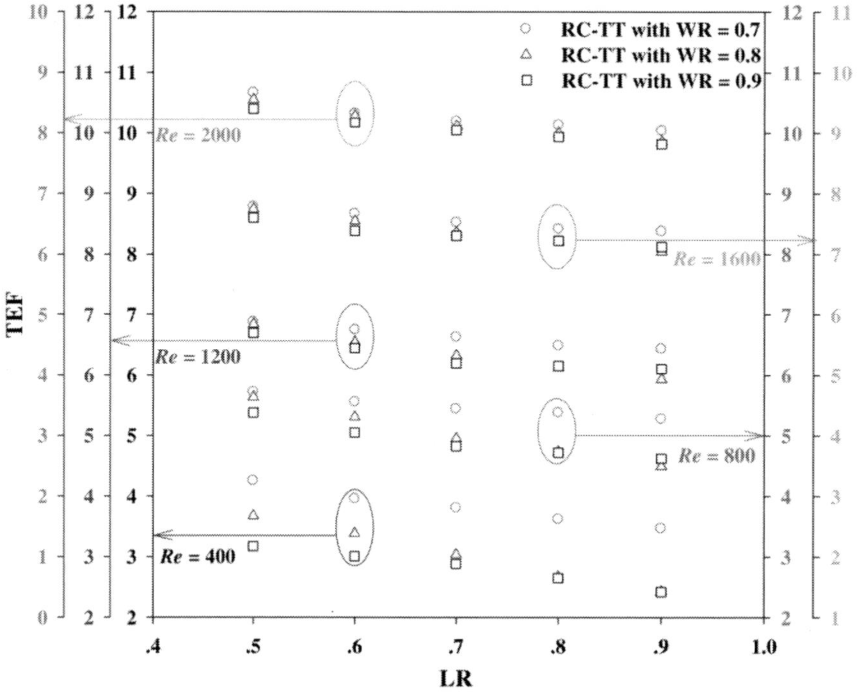

Figure 8. Influence of perforated length to tape width ratio (*LR*) and perforated width to tape width ratio (*WR*) on thermal enhancement factor (*TEF*) at different Reynolds numbers (*Re*).

Performance evaluation

Figure 8 presents the thermal enhancement factor (*TEF*) for water flowing in the tube with rectangular-cut twisted tape inserts. For all cases, thermal enhancement factor substantially increases with increasing Reynolds number. All tubes with rectangular-cut twisted tape swirl generators yield thermal enhancement factors above unity at the highest Reynolds number (*Re* = 2000). Among the studied rectangular-cut twisted tape swirl generators, the one with the smallest *WR* of 0.7 and the smallest *LR* of 0.5 offers the maximum thermal performance factor of 8.66.

Conclusions

Heat transfer enhancement, turbulence flow field, fluid temperature, local Nusselt number, pressure loss and thermal performance characteristics of the tubes equipped with fifteen different rectangular-cut twisted tapes possessing perforated width to tape width ratios ($WR = w/W$) of 0.7, 0.8 and 0.9, and perforated length to tape width ratios ($LR = L/W$) of 0.5, 0.6, 0.7, 0.8 and 0.9 are numerically investigated. Evidently, the insertion of twisted tapes leads to the promoting velocity and reducing fluid temperature near wall regions, improving fluid mixing between core and near wall regions and thus thinning thermal boundary layers as compared to those in the plain tube. At a given Reynolds number, heat transfer, friction loss and thermal enhancement factor (*TEF*) increase with decreasing perforated width to tape width ratio (*WR*) perforated length to tape width ratio (*LR*). Among the studied rectangular-cut twisted tape swirl generators, the one with the smallest *WR* of 0.7 and the smallest *LR* of 0.5 offers the maximum thermal performance factor of 8.66.

References

1. He, Y., Liu, L., Li, P. and Ma, L. (2018), "Experimental Study on Heat Transfer Enhancement Characteristics of Tube with Cross Hollow Twisted Tape Inserts," Applied Thermal Engineering, Vol. 131, pp. 743-749.
2. Thianpong, C., Eiamsa-ard, P., Promvonge, P. and Eiamsa-ard, S. (2012), "Effect of perforated twisted-tapes with parallel wings on heat transfer enhancement in a heat exchanger tube," Energy Procedia, Vol. 14, pp. 1117-1123.
3. Thianpong, C., Eiamsa-ard P. and Eiamsa-ard, S. (2012), "Heat transfer and thermal performance characteristics of heat exchanger tube fitted with perforated twisted-tapes," Heat and Mass Transfer/Waerme- und Stoffuebertragung, Vol. 48, pp. 881-892.
4. Nanan, K., Thianpong, C., Promvonge, P. and Eiamsa-ard, S. (2014), "Investigation of heat transfer enhancement by perforated helical twisted-tapes," International Communications in Heat and Mass Transfer, Vol. 52, pp. 106-112.
5. Eiamsa-ard, S. and Promvonge, P. (2011), "Influence of double-sided delta-wing tape insert with alternate-axes on flow and heat transfer characteristics in a heat exchanger tube," Chinese Journal of Chemical Engineering, Vol. 19, pp. 410-423.
6. Hasanpour, A., Farhadi, M. and Sedighi, K. (2016), "Experimental heat transfer and pressure drop study on typical, perforated, V-cut and U-cut twisted tapes in a helically corrugated heat exchanger," International Communications in Heat and Mass Transfer, Vol. 71, pp. 126-136.
7. Patankar, S.V., (1980). "Numerical heat transfer and fluid flow," McGraw-Hill, New York.
8. Incropera, F.P., Dewitt, P.D., Bergman, T.L. and Lavine, A.S. (2006), "Fundamentals of Heat and Mass Transfer," John-Wiley & Sons, New York.
9. Manglik, R.M. and Bergles, A.E., (1993), "Heat transfer and pressure drop correlations for twisted-tape inserts in isothermal tubes: Part I-Laminar flows," Transactions of the ASME Journal of Heat Transfer, Vol. 115, pp. 881-889.

540503 Concept to Commercialization: Energy efficient & Eco-friendly Anode Grade Coker Technology

Abstract:

IndianOil R&D has developed the process know-how of 'Energy Efficient & Eco-friendly Anode Grade Coker Technology' with superior features compared to conventional Anode Grade Coker technology. This technology was successfully commercialized in 600 TMTPA Capacity Coker unit at Barauni Refinery (BR), India, which converts heavy hydrocarbon residues into higher distillates with lower coke yield meeting 'Anode Grade' specifications.

The major merit of this technology is the production of 'Anode Grade Coke' employing a low recycle ratio, similar to that used in contemporary 'Fuel Grade Coke' producing Coker units. This technology deployment by IndianOil is a landmark achievement for petroleum refining industry. Earlier, Coker unit at BR was being operated at a high recycle ratio in the range of 70-80% producing lower distillate and higher coke yield with an objective to produce 'Anode Grade Coke'. After the incorporation of this new technology, the refinery is now able to produce 'Anode Grade Coke' at a low recycle ratio (5-10%) with much superior yield pattern in terms higher distillate and lower coke. Besides improving the unit margin, it also reduces the energy consumption and thus, the carbon foot prints. The commercial unit was smoothly commissioned in January 2017 and Performance Guarantee Test Run successfully completed in March 2017 meeting all the guarantee parameters. Significant benefits due to implementation of the technology are as below,

- Reduction of coke yield by ~ 4.4 wt% & increase in distillate yield by ~ 4 wt%
- Savings in energy by ~ 37%
- Significant reduction in pollutants like CO_2, SO_X, NO_X, carcinogenic hydrocarbons and dust
- Enhanced safety & reliability
- GRM improvement by ~ 25 Million USD/Year

Successful implementation of this technology is a momentous milestone in refining industry and its adoption in refineries would improve the refining margin while maintaining the Anode grade coke quality, wherever produced.

This paper describes the journey from concept to commercialization of 'Energy Efficient & Eco-friendly Anode Grade Coker Technology' in BR Coker.

540377 Quantified Investigation of Coal Ash Fusion Behavior for Gasifier Design and Operation

Abstract:

Coal gasification is regarded as the primary technology for clean coal utilization in large scale. Through coal gasification, chemicals and liquids can be supplied instead of only heat by coal combustion. The entrained flow gasifier is the priority choice for coal to chemicals and liquids because it is running at high temperature and pressure. In the high temperature gasification process, coal matrix transforms into syngas, but minerals go through temperature increasing and decreasing into ash and slag. Ash chemistry is proved to be more important than coal matrix at high temperature gasification. Ash fusion behavior is used as the first criteria for coal selection for gasifier. However, the traditional ash fusion temperatures used for 100 years are lack of scientific definition and of wide error range, so the application of ash fusion temperatures are more empirical. In this work, our used thermomechanical analyzer, microscopy coupled with heating stage and DSC to investigate the ash fusion behavior. The ash fusion process is divided into three stages according to the fusion rate. Deformation temperature and flow temperature is defined with physical meaning. Based on the TMA curve and thermodynamic calculation, the relation between ash composition and sintering rate is established. Meanwhile, with the help of molecular simulation, we established the new method to calculate the liquidus temperature of coal ash slag. The results could be used for gasifier operation and design.

523729 Conversion from CO_2-Containing Flue Gas to Electrocatalysts

Abstract:

Management of flue gas, which contains about 15% of CO_2 and an excess amount of N_2, is an important issue to diminish CO_2 in atmosphere. Herein, CO_2 contained in flue gas was directly converted into â€˜boron and nitrogen co-doped porous carbon materials (FlueBNPC)â€™ at 500 °C and atmospheric pressure without any additional separation process. The synthesized FlueBNPC was applied as an electrocatalyst for oxygen reduction reaction (ORR) in fuel cell cathode. Boron and nitrogen in the carbon network, which were doped simultaneously from reducing agent ($NaBH_4$) and flue gas, acted as active sites for electrochemical reaction. The synthesized Flue-BNPC showed better electrocatalytic activity toward ORR compared to pristine BPC which was converted from pure CO_2. This enhancement was due to the synergetic effect of B-N bonding combined with oxygen atoms in the carbon network, which was investigated not only by experimental characterization, but also by DFT calculations. The results demonstrate that flue gas itself can be an effective precursor for synthesis of carbon materials, insertion of B/N atoms into carbon lattices, and finally, production of efficient carbon-based electrocatalysts.

520599 Composite Electrolytes of Pyrrolidone-Derivatives-PEO Enable to Enhance Performance of All Solid State Lithium-Ion Batteries

Abstract:

All-solid-state batteries with $LiFePO_4$ cathodes and polyethylene oxide (PEO) based electrolytes are optimized with organic polymer synthesized from allyl amine and tartaric acid which show better properties including ionic conductivity, electrochemical window, charge-discharge capacity and rate performance. The rate of PEO (= 4,000,000) and lithium salt (Lithium bis(trifluoromethanesulphonyl)imide, LiTFSI) is 75:20 (w/w) and polymers of different molecular weight are tested at a wide temperature range from 27 °C (ambient temperature) to 120 °C. Electrolytes with larger amount (weighting more than half of PEO) of low-molecular-weight polymer show conductivity of above 10^{-4} S cm^{-1}, which is similar to typical gel-polymer-electrolyte. Electrolytes which are made from high-molecular-weight dopant, PEO and LiTFSI through ball milling show better conductivity (5.4×10^{-6} S cm^{-1}) as well as better strength than electrolytes without dopant (1.3×10^{-6} S cm^{-1}) at room temperature. And a sensitive dependence on temperature increasing is also observed constructing to the sample without dopant. Stability below 120 °C and an electrochemical window of 5.4 V at 80 °C also indicate promising use of this composite electrolyte in solid state lithium ion batteries.

533758 Direct Power Generation from Reed Biochar in a Direct Carbon Fuel Cell

Abstract:

A direct carbon fuel cell (DCFC) is a promising power generation device converting the chemical energy stored in solid carbon into electrical energy through a electrochemical process with a high energy efficiency and low environmental impacts. The fuels of DCFCs originate from various carbon sources, which is a unique characteristic compared to the conventional fuel cell. The feasibility of biochar obtained from the pyrolysis of reed as the fuel of DCFCs is investigated in the present study. At 750 ^{o}C, the cell fed with reed biochar using a $Ce_{0.8}Sm_{0.2}O_{1.9}$ (SDC)-carbonate composite electrolyte exhibits a maximum power density of 351 mW cm^{-2}. The high performance of the cell is attributed to the metal species in the biochar, the low graphitization degree and the high amount of oxygen-containing functional groups of the biochar.

533792 A $LaNi_{0.9}Co_{0.1}O_3$ Coated $Ce_{0.8}Sm_{0.2}O_{1.9}$ Composite Anode for Solid Oxide Fuel Cells Fed with Methanol

Abstract:

Solid oxide fuel cells (SOFCs) are promising devices those convert chemical energy in fuels directly into electricity with a high efficiency. The most common anode of SOFCs consists of metallic Ni as the catalyst and electronic conductor and a ceramic phase as the oxygen ionic conductor. The content of Ni is usually higher than 45 vol% to achieve a sufficient electronic conductivity. However, such a high amount of Ni suffers from not only aggregation at a high temperature, which results in the reduction of active sizes and the catalytic activity, but also serious carbon deposition with hydrocarbons as fuels. Here $LaNi_{0.9}Co_{0.1}O_3$ (LNC)-$Ce_{0.8}Sm_{0.2}O_{1.9}$ (SDC) anodes are prepared and investigated as anode materials of SOFC fed with methanol.

The composite anode is synthesized by impregnation. Nanoparticles of Ni-Co alloy are derived from LNC after reduction. The composite anode shows a sufficient electrical conductivity when the weight ratio of Ni to SDC reaches 0.1, which is much lower than the amount of Ni required in conventional Ni-SDC anodes. The aggregation of Ni-Co nanoparticles in the reduced anode is suppressed by the oxide supporter, resulting in an improved catalytic activity. The single cell with LNC-SDC composite anode and SDC-carbonate composite electrolyte exhibits a maximum power density of 872 mW cm^{-2} at 700 ^{o}C with gasified methanol as the fuel, much higher than that of the cell with a NiCo-SDC anode under the same condition. The amount of carbon deposited on the anode in methanol atmosphere decreases with the decline of the LNC content. The cell with the 10LNC-SDC anode exhibits a stable performance during a 10 h discharge period.

Effect of Geometry gas Flow Rates, and Oxygen Concentration on the performance of Anode-Supported Planar Solid Oxide Fuel Cells

Nayan Biswas[1,*], Deepra Bhattacharya[2], Jayanta Mukhopadhyay[2], Rajendra Nath Basu[2], Prasanta Kumar Das[1]

Abstract

The modern era is seeking highly efficient power generation systems which have minimum carbon footprints. The Solid Oxide Fuel Cell (SOFC) is an example of such a kind of device that generates power from the fuels (Hydrogen, Hydrocarbons etc.) in the presence of oxidants (pure oxygen, air) without any direct combustion. In the present study, the power characteristics of the serpentine geometry with two 180 bends is modelled and analysed to investigate its characteristics at various operating conditions. The performance of SOFCs is analysed varying the oxygen concentration of the oxidants ranging from pure oxygen to air. The study depicts the enhancement of performance with increasing flow rate is significant up to a limit and further the performance is not enhanced significantly with the increment of flow rates for the present geometry. Further to this, a study of geometric variation is also carried out to evaluate the dependency of cell performance over various geometric dimensions of the cell. The investigation can give a direction for the optimization of flow rates of cathode along with their composition

Keywords: SOFC modelling, Serpentine channel, Power density, overpotential.

1. Introduction

The modern era is seeking highly efficient power generation systems which have minimum carbon footprints. The Solid Oxide Fuel Cell (SOFC) is an example of such a kind of device that generates power from the fuels (Hydrogen, Hydrocarbons etc.) in the presence of oxidants (pure oxygen, air) without any direct combustion. Amongst all other fuel cells Solid Oxide Fuel Cells (SOFCs) are more beneficial due to its high temperature, fuel flexibility, and inexpensive catalysts [1]. The working principle along with the electrochemical reactions are shown using a schematic diagram of an SOFC in Figure 1.

An unit cell of an SOFC consists of six major components; two porous electrodes, named as cathode and anode; a solid dense electrolyte, is sandwiched between the two electrodes.There are two fluid flow channels adjacent to each electrode. The fluid flow paths depend on the shape of interconnects. Besides this, the metallic interconnect collects current from the electrodes. The hydrogen gas(H_2) is introduced in anode channel.

*Corresponding author

[1]Department of Mechanical Engineering, Indian Institute of Technology Kharagpur

[2]Fuel Cell and Battery Division, Central Glass and Ceramic Research Institute

September 12, 2018

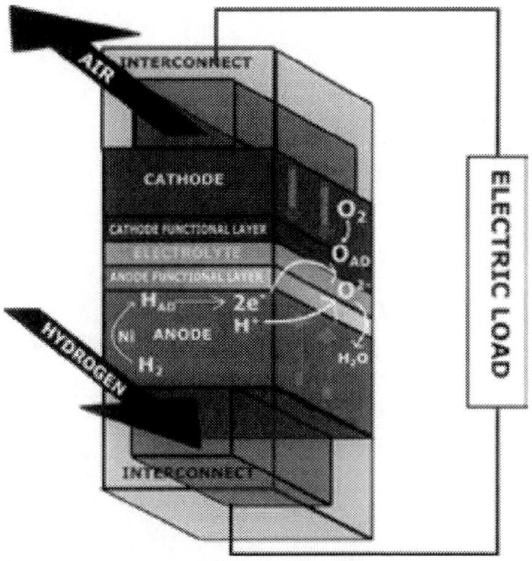

Fig. 1. Schematic Diagram of a Solid Oxide Fuel Cell showing electrocatalytic reactions [2]

The H_2 gas is diffused through porous anode and reach to the overlapping boundary of anode and electrolyte, popularly known as Triple phase boundary (TPB), Where they oxidized and produce H^+ ions. In the cathode side, the oxygen is diffused through porous cathode and produce O^{2-} ions in the cathode-electrolyte TPB. The O^{2-} ions are conducted through electrolyte and reached to anode-electrolyte TPB and reacts with H^+ ions to produce water. In the present study, a thorough numerical simulations have been carried out to understand the effect of cell geometry on the performance of an SOFC. A parametric study is also done to study the effect of gas flow rate as well as gas flow composition to optimise the operating condition of an anode supported SOFC using a commercial finite element based CFD software COMSOL Multiphysics software.

2. Methodology

In the present literature, a 3-D isothermal model of planar SOFC is developed. The model is studied using different complex sets of algebraic/differential equations for multiple physics at particular domains associated with suitable boundary conditions. All the equations are discretized and solved using commercial CFD software COMSOL Multiphysics [3]. The mesh refinement study and validation with the experimental data is already provided in our earlier article [2]. In the present study, the diffusional properties, material properties are taken as dependent of temperature as well as pressure. The study is also considered all the overpotentials, i.e., activation, concentration and ohmic overpotential, associated with an SOFC [2].

2.1. Model setup and Geometric dimensions

The current planar serpentine channel SOFC is composed of Nickel-doped Yttria Stabilised (Ni-YSZ) anode, 8 mol% Yttria Stabilished Zirconia (8-YSZ) electrolyte and Strontium-doped Lanthanam Manganite cathode and a metalic interconnect. In the present model, the solid electrolyte is an electronic insulator, but O^{2-} ions can conduct through electrolyte.

The geometric dimensions are given in the Table 1.

Table 1. Modelling Parameters

Parameter(Unit)	Symbol	Value/Expression	Source
Anode Thickness (µm)	H_{anode}	150	
Electrolyte Thickness (µm)	$H_{electrolyte}$	30	
Cathode Thickness (µm)	$H_{cathode}$	50	
Anode Flow Channel Height (mm)	H_{aChan}	1	
Cathode Flow Channel Height (mm)	H_{cChan}	1.3	
Interconnect Thickness (mm)	H_{int}	$1.5 \times H_{electrode}$	
Specific Surface Area (m^{-1})	S_g	1.2E+5	[4]
Fuel/Air Inlet Temperature (K)	T_{in}	1073	
Ambient Temperature (K)	T_{amb}	1073	
Ambient Pressure (Pa)	P_{atm}	101325	

2.2. Governing equations and boundary conditions

SOFC modelling associated with number of complex equations with various physics, which are solved simultaneously with particular boundary conditions. Flow in cathode and anode gas channels are modelled using weakly compressible Navier Stokes Equation. The subsurface flow in the porous electrodes are narrated using Brinkman equation. The flow rate in the cathode and anode inlet is given as an inlet boundary conditions and the outlet of the channel is assumed to be at ambient pressure. No slip boundary condition is provided at the adjacent of the solid electrolyte wall. A general energy equation is used to depicts the temperature variation in the cell. The gas inlet temperature of fuels and oxidant are at 1073 K and is given as the inlet boundary condition for the heat transfer. Mass transport is governed by bulk motion of fluid in the channels as well as by diffusion in the porous electrodes. The Maxwell-Stefan diffusion model is implemented to study mass transport in the porous medium. The electro-chemistry of the cell is modelled using concentration dependent Butler-Volmer equation. The cathode-interconnect surface is considered at the cell operating voltage; whereas, the anode-interconnect surface is at

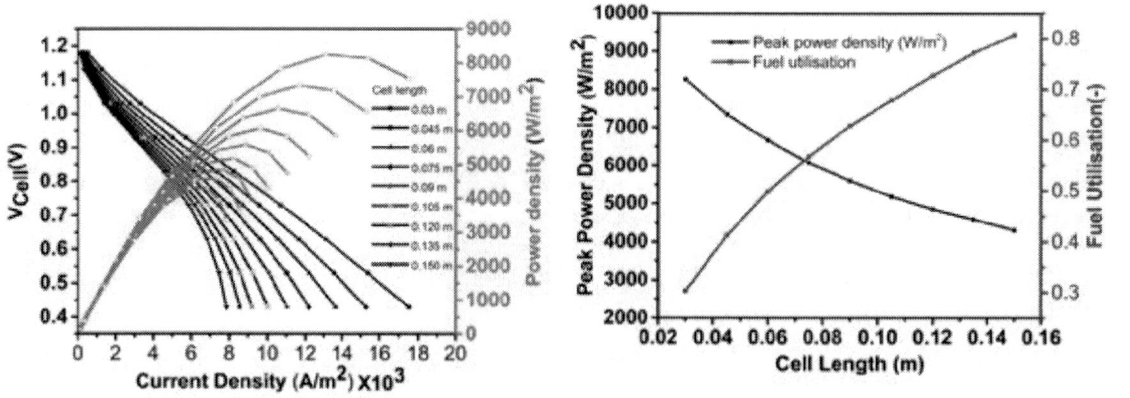

(a) VI characteristics

(b) Variation of Peak power and fuel utilisation

Fig. 2. Effect of Cell length

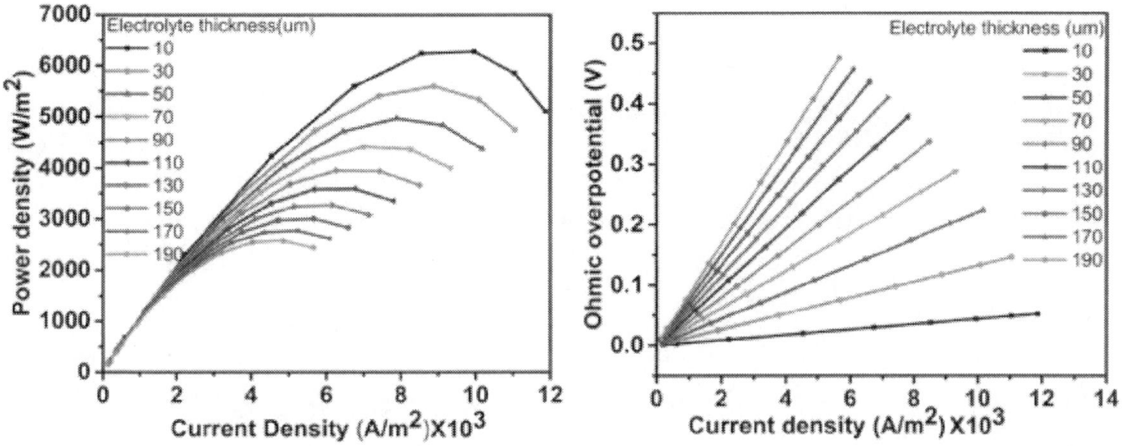

(a) VI characteristics

(b) Variation of electrolyte overpotential

Fig. 3. Variation of electrolyte thickness

ground (0 V). Periodic boundary conditions are applied at all suitable boundaries parallel to the gas flow.

The model is uniformly meshed using $\sim 10^6$ cuboidal elements customized to be finer near the reacting interfaces. A direct PARDISO solver is used to solve the equations with a relative tolerance of 10^{-5}.

3. Results

The performance of the numerical study by varying the parameter to analyse the effect of design and other operating parameters so that the performance of an SOFC could be improved and the operating conditions can be optimised.

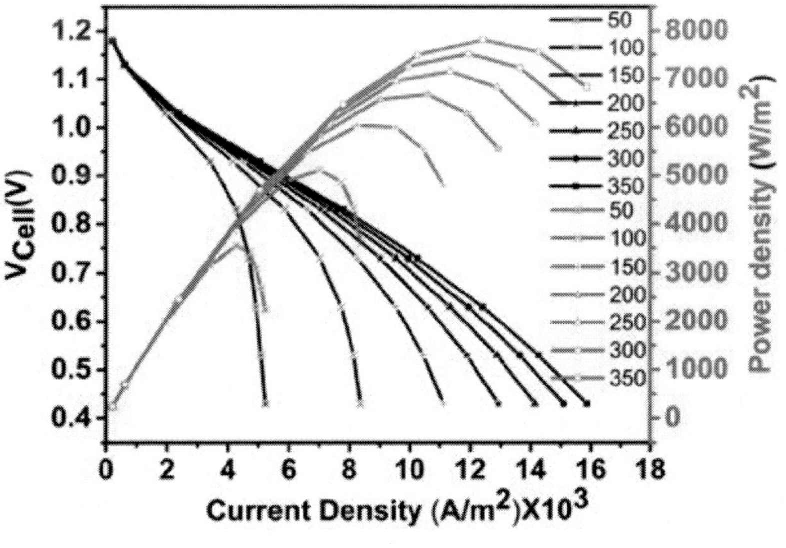

Fig. 4. variation of cathode flow rate

3.1. Geometric variation

3.1.1. Effect of the cell length

Figure 2 depicts about the performance of the cell due to variation of the cell length. The length of the cell is varied from 30 mm 150 mm. Figure 2a shows the electrochemical characteristics of SOFC for each cell length. It is observed that as the length of the cell is increased the current density as well as the power density is decreased. When the length of the cell is increased, the concentration of the H_2 in the anode is decreased towards the outlet of the cell. As a result, the current density and power density is decreased with increment of cell length.

Figure 2b shows the variation of peak power density with increasing cell length, and corresponding variation in fuel utilisation. It is already discussed that due to decrement of H_2 concentration, the power density is reduced. However, as the cell length is increased, the active area for reaction as well as the residence time is increased. As a result, the corresponding fuel utilisation is raised with increment of cell length.

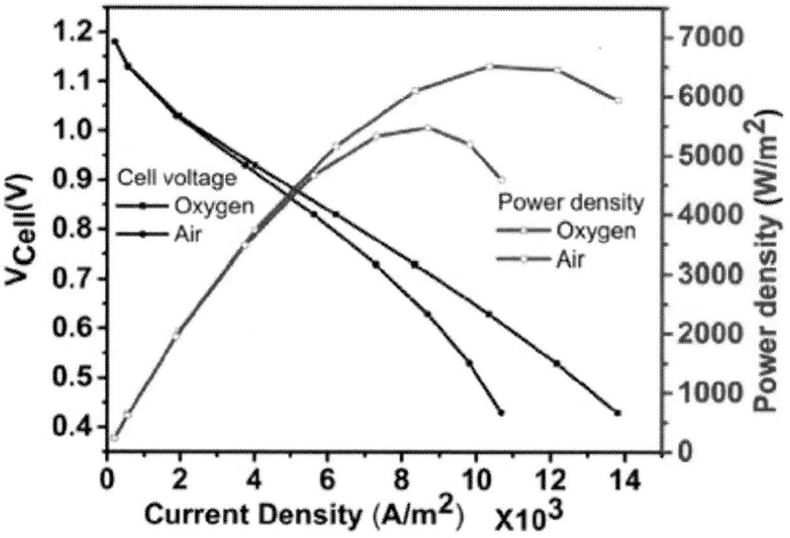

Fig. 5. Variation of cathode gas composition

3.1.2. Variation of electrolyte thickness

Figure 3 shows the effect of electrolyte thickness on performance of an SOFC. The thickness of the electrolyte is varied from 10 µm to 190 µm. Figure 3a shows the variation of power density with change of electrolyte thickness. The figure depicts that as the electrolyte thickness is increased, the power density of the cell is decreased. Figure 3b shows the variation of ohmic loss of electrolyte with change of current densities at different electrolyte thickness. As the electrolyte thickness of the model is increased, the ohmic overpotential associated with the cell is also raised. As a consequence, the power density is decreased with enhancement of electrolyte thickness. The ohmic overpotential of anode and cathode can be neglected with comparison to electrolyte ohmic loss. The influences are highest at low electrolyte thickness and weaken as the thickness of the electrolyte is increased.

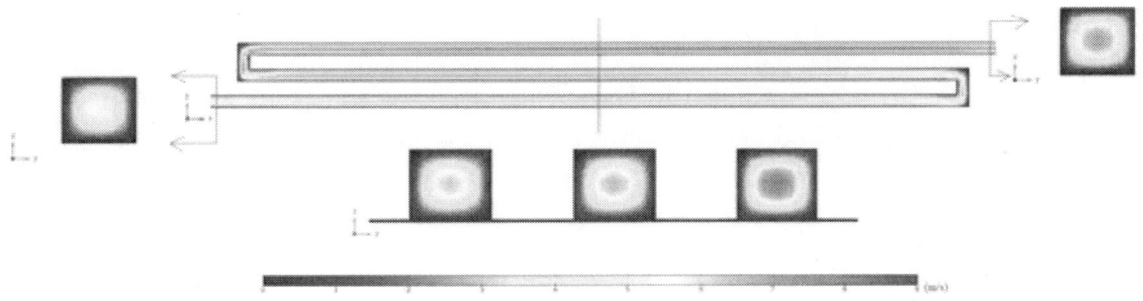

Fig. 6. Spatial distribution of velocity at Cathode channel

3.2. variation of cathode flow rate

SOFCs are operated generally on stoichiometric air fuel ratio. Figure 4 shows the performance of an SOFC with variation of air flow rate from 50 SCCM to 350 SCCM in the cathode channel, keeping the fuel flow rate constant at 75 SCCM (Stoichiometric air flow rate 175 SCCM). the figure depicts that air flow rate at less than stoichiometry, leads to a significant decrement in performance. The enhancement in power density due to increment of air flow rate, is higher at lower flow rate and weaken towards high flow rate.

3.3. Variation of Cathode gas composition

Figure 5 depicts the performance of an SOFC with alteration of oxidant in cathode side from air to pure oxygen. The curve clearly shows that as the oxidant is altered with oxygen instead of air, the maximum power density is raised by 30%. The VI characteristics depicts that the variation is more at higher current density; which implicates the concentration loss due to the presence of Nitrogen gas (N_2) in the air.

3.4. Spatial distribution of Velocity

3.4.1. Cathode channel

figure 6 depicts the cathodic velocity profile for a serpentine channel SOFC. Due to the no slip boundary condition adjacent to the solid interconnect wall, the fluid is in rest. the velocity of cathode flow is higher at the inlet and it is reduced with advancement of flow, as the flow rate is reduced due to utilisation of oxygen.

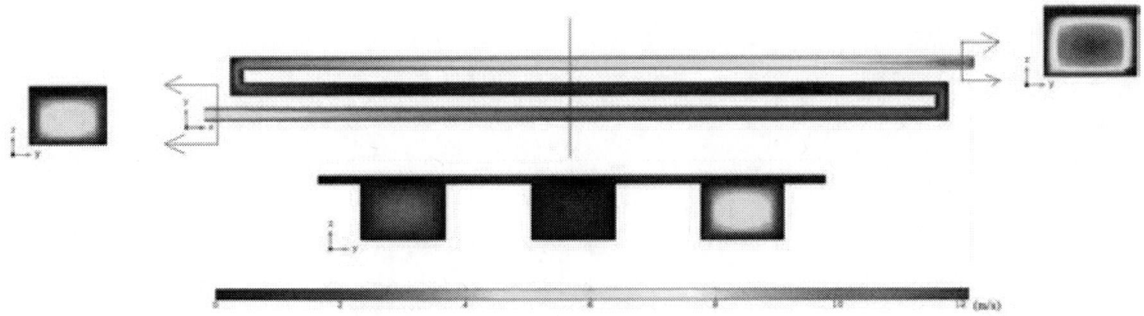

Fig. 7. Spatial distribution of velocity at Anode channel

3.4.2. Anode channel

The velocity profile in the anode side shows an interesting variation. The velocity is high at inlet and it is gradually decreased in the first segment due to high fuel utilisation. The decrement is being continued in the second segment too. but, in the last segment the velocity increases steadily due to evaluation of water.

4. Conclusion

It has been concluded that the performance of an SOFC is highly sensitive to its geometry as well as at it's operating condition. A number of parametric variation is conducted to optimise the operating condition of a serpentine SOFC. It is found that increasing the length of the cell, the fuel utilisation is increased despite of it's decrement in power density. The study also reveals that increasing the electrolyte thickness enhance the ohmic polarisation; as a consequence, the performance of the cell is decreased. Further, the investigation reveals the performance dependency of the cell on oxidant concentration and it has been found that the cell performance is increased as the oxidant is changed from air to pure oxygen.

Acknowledgements

The infrastructure and the technical support to continue the research activities from the Director, CSIR – CGCRI and Director, IIT – Kharagpur is highly acknowledged by the authors. The financial support from CSIR – NMITLI is also acknowledged.

References

[1] N. Minh, Solid oxide fuel cell technology-features and applications, Solid State Ionics 174 (2004) 271–277.

[2] D. Bhattacharya, J. Mukhopadhyay, N. Biswas, R. N. Basu, P. K. Das, Performance evaluation of different bipolar plate designs of 3d planar anode-supported {SOFCs}, International Journal of Heat and Mass Transfer 123 (2018) 382 – 396.

[3] COMSOL, COMSOL 5.3 Documentation, 2017.

[4] M. Nerat, D. Juricic, A comprehensive 3-D modeling of a single planar solid oxide fuel cell, International Journal of Hydrogen Energy 41 (2016) 3613–3627.

526181 Experimental Optimization of Design Parameters of Cylindrical PEM Fuel Cell and Diagnosis of Its Performance Degradation

Abstract:

Proton Exchange Membrane (PEM) fuel cells have the potential to replace secondary batteries in portable applications. However, the heavy subsystems, expensive components like graphite plates and titanium end plates are the barriers for its commercialization in this sector. This has led to the development of air-breathing cylindrical PEM (c-PEM) fuel cells, which eliminate the need for graphite plates and titanium endplates. The preliminary design of a air-breathing cylindrical fuel cell delivers a performance of 300 mA/cm^2 @ 0.6 V. Since the design parameters of this cell have not been optimized, there is still considerable scope for improving performance.

One of the design parameters that can affect the cylindrical cell performance is cathode open area. Larger open area for cathode will increase the path length for electrons in gas diffusion layer and the smaller open area will increase the mass transport loss by reducing the oxygen access to the cathode. Decreasing the cathode open area will also increase the rib thickness of cathode current collector, which will lead to the improper distribution of oxygen over the cathode. The orientation of cathode opening is another parameter which can influence the cell performance. Horizontal slots in the cathode will obstruct the free fall of water droplets produced on the cathode. The generated water molecules can stick to the rib of the cathode current collector and prevent the oxygen molecules from reaching the cathode catalyst layer. On the other hand, vertical slots in the cathode current collector will provide less number of sites for water molecules to stick to its rib. The geometry of the cathode opening can also affect the cell performance because the hydraulic diameter of the cathode opening is dependent on its geometry. The first part of the talk will focus on the optimization of c-PEM fuel cell design parameters such as cathode open area, orientation and geometry of cathode opening.

During the initial testing of the cell, it was observed that the cell performance degrades over time even when the cell is not in operation. The reason for this performance degradation can be due to the cathode in c-PEM fuel cells being always open to atmosphere, unlike planar fuel cells. This can give rise to two problems, one is the higher rate of water evaporation from the membrane and the other is the corrosion of carbon support. Loss of water from the membrane will increase the ohmic resistance of the cell and the carbon support corrosion can decrease the electrochemically active surface (ECSA) of the electrode. The second part of this talk will be dedicated to the development of a test protocol incorporating impedance and ECSA measurements to identify the reason behind this performance degradation. A solution strategy will also be developed to improve the durability of c-PEM fuel cells.

Organic Rankine Cycle Waste Heat Recovery System to Cool the Data Center

ID# 534834

Russell C. Tipton [a,b,], M. Toufiq Reza [a,c],*
[a] Department of Mechanical Engineering, 1 Ohio University, Athens, OH 45701, USA
[b] Vertiv, 1050 Dearborn Drive, Columbus, OH 43085, USA
[c] Institute for Sustainable Energy and the Environment, 350 W State St., Athens, OH 45701, USA
[*] Corresponding author: Tel.: +1 614 314 0605, E-mail addresses: rt448415@ohio.edu, russ.tipton@vertivco.com (R.C. Tipton)

Abstract

An opportunity exists to improve the process of removing waste heat from IT equipment in data centers through use of a heat recovery process. The information and communication technology industries (ICT) consume ~2% of globe's energy which translates to ~4% of electricity. Nearly all the electrical input power results in waste heat. Demand for increased compute speed and capacity while minimizing operation latency has driven processor heat density beyond what can be cooled with air. Processors operate successfully at temperatures up to 95°C. Direct liquid cooling (DLC) can capture the waste heat at 85°C making it available for heat recovery between at 75°C and 80°C. Smaller data centers (aka Edge) are less than 500kWe IT load. These facilities utilize traditional cooling techniques that consume energy at an average rate of ~40% of the IT load. Nearly 3,000,000 sites exist globally.

ORC has proven effective in heat recovery applications with low temperature differentials. Average ambient temperature of ~14°C provides 60°C to 65°C driving force for the ORC system. At 50% of ideal efficiency and allowing for heat rejection power, the ORC can deliver ~7% of the input waste heat as high-quality energy. The primary technology gap in ORC is the expander. Various research projects have found scroll machines to be effective expanders. A focus of this research effort is to design and test scroll expanders producing a design that may be commercialized.

The HVACR industry has developed efficient long life scroll compressors. The basic scroll device is reversible. Identifying and developing the design changes to convert commercial scroll compressor requiring electrical power input to a scroll expander producing electrical power output is required to enable this process improvement. In simple terms, an ORC heat recovery system is an HVACR vapor compression refrigeration system operating in reverse. The components of the two systems are comparable and expected to be similar in cost. The process improvement results in reduction of energy required to remove the waste heat from ~40% of IT load to a ~7% net supply of electrical energy. Total data center energy usage can be reduced by ~32%.

Introduction

ICT consumes ~4% of globe's electricity

A conservative estimation for 2017 puts the world's Information-Communications-Technologies (ICT) ecosystem consumption at ~4% of electricity which produces ~2% of CO_2 emissions. [1, 2]

1. Giorgos Fagas, John P. Gallagher, Luca Gammaitoni and Douglas J. Paul (March 22nd 2017). Energy Challenges for ICT, ICT - Energy Concepts for Energy Efficiency and Sustainability Giorgos Fagas, IntechOpen, DOI: 10.5772/66678. Available from: https://www.intechopen.com/books/ict-energy-concepts-for-energy-efficiency-and-sustainability/energy-challenges-for-ict
2. Lotfi Belkhir, Ahmed Elmeligi. Assessing ICT global emissions footprint: Trends to 2040 & recommendations. Journal of Cleaner Production. Volume 177, 10 March 2018, Pages 448-463.

The target application is heat recovery from ICT equipment. The operating temperature limits of the ICT electronics are below 85°C. [1].

PUE of 400kW IT facility / Target single system capacity

The landscape of the ICT world is shown in Figure 1. As shown, Network Closets and Server Rooms are less than 5,000 ft2, contain between one and fifty servers, and roughly 2,800,000 of these sites exist. [2] The need for high speed streaming of content, collection of large volume local generation of data, and high-performance computing drives the need for these physically-close-to-application data centers. [3, 4, 5, 6, 7, 8]

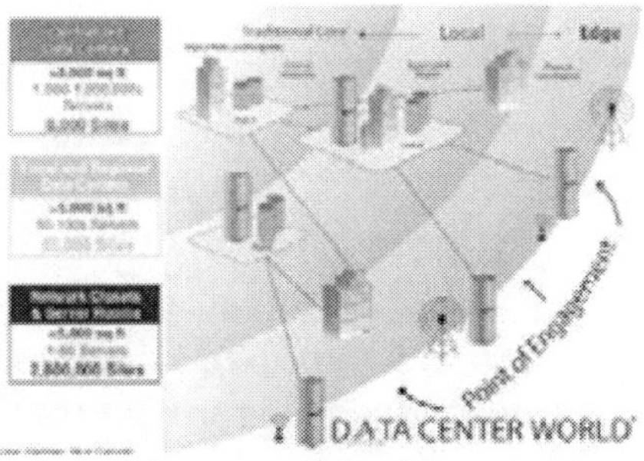

Figure 1. Architecture of the data center world. [4]

Figure 2. shows average ICT industry power usage effectiveness as of March 2012 based on a 400kW IT facility. Specifically, IT load is 400kW; building and IT loss is 24.1kW; lighting and general receptacle load is 1kW, and cooling requires 196kW to move the waste heat. Using this data, resulting power usage effectiveness (PUE), ratio of total facility power and IT power, is 1.6. [9] Thus, cooling accounts for 32% of data center power usage.

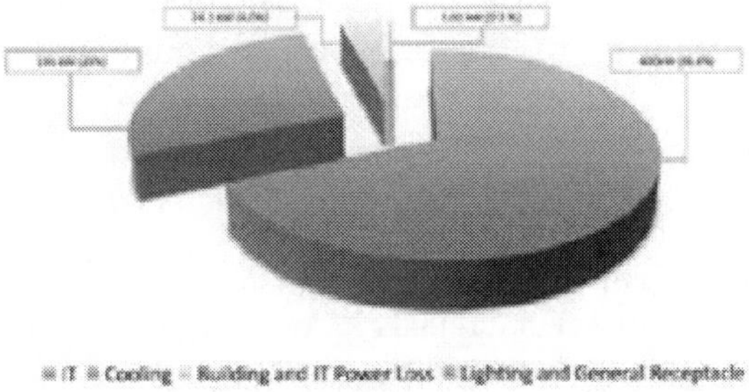

Figure 2. Power Usage Effectiveness – March 2012 [9]

1. Khosrow Ebrahimi, Gerard F. Jones, Amy S. Fleischer. A review of data center cooling technology, operating conditions and the corresponding low-grade waste heat recovery opportunities. Renewable and Sustainable Energy Reviews 31 (2014) 622–638.
2. Jack Pouchet. Five Key Emerging Trends Impacting Data Centers in 2016. Data Center World. AFCOM. Sep-2016.
3. Datacom equipment power trends and cooling applications. 2nd ed. Atlanta (GA,USA): ASHRAE; 2005.
4. Yevgeniy Sverdlik. Deep Learning Driving Up Data Center Power Density. Data Center Knowledge. http://www.datacenterknowledge.com/archives/2017/03/27/deep-learning-driving-up-data-center-power-density. Accessed 24-Sep-2017.
5. Alex Carroll. Data Center Power Density. https://lifelinedatacenters.com/data-center/data-center-power-density/. Accessed 24-Sep-2017.
6. John Musilli, Brad Ellison. Facilities Design for High-density Data Centers. Intel Corporation, 2012.
7. Robert L. Mitchell. Data center density hits the wall. Computerworld. 21-Jan-2010. https://www.computerworld.com/article/2522601/it-management/data-center-density-hits-the-wall. Accessed 24-Jan-2017.
8. Jeff Clark. Raising Data Center Power Density. The Data Center Journal. 24-Oct-2013.
9. Data Center Utilization Report – March 2012. Information Technology Division. A U.S. Department of Energy National Laboratory Operated by the University of California.

Network Closets and Server Rooms are the focus heat recovery system for system sizing purposes in this work. Ebrahimi et. al. state that the total power consumption of servers is typically between 300W and 400W. Further, servers can be enclosed in standardized metal racks which can handle up to 42 servers each. This puts power consumption of a fully populated rack at ~15kW. However, expectations are that a single rack filled with extreme density communication equipment may consume 60kW and when filled with extreme density computer servers, 35kW. [1] A reasonable target heat recovery system capacity is 50 kW thermal power input.

Temperature differential in ICT application

The temperature profile is a solid reference point to begin the design of a thermodynamic cycle. A maximum 85°C junction temperature can achieve effective operation of microprocessors. Small temperature difference between heat source and waste heat stream is produced by direct liquid cooling (DLC) of the IT equipment. [1] Weather maximum and minimum temperature design conditions for populated areas of continental USA range from 45°C to -20°C while the average temperature is between 11°C and 25°C depending on location. [2] A broad efficient system operating range is desired when the operating conditions vary. For this ICT application, design conditions are set at 85°C heat source temperature and 18°C heat sink temperature. Separation of ORC working fluid from IT equipment cooling fluid is prudent. A 10°C total differential provides 5°C for heat transfer from IT heat source to IT coolant and 5°C for heat transfer from IT coolant to ORC working fluid. A 5°C differential for heat transfer from ORC to heat sink is a reasonable condenser design goal. These provide design conditions for ORC operating at TH=75°C and TL=23°C producing a 52°C design temperature differential.

Organic Rankine Cycle

Several thermodynamic cycles for low-grade heat-to-power conversion have been investigated, ORC is the most commercially developed with both large and small power plants in operation. [3] ORC, compared with other heat engine cycles, exhibits high waste energy utilization. [4] Organic Rankine cycle characteristics include: simple structure, high reliability, and easy maintenance. [5]

1. Khosrow Ebrahimi, Gerard F. Jones, Amy S. Fleischer. A review of data center cooling technology, operating conditions and the corresponding low-grade waste heat recovery opportunities. Renewable and Sustainable Energy Reviews 31 (2014) 622–638.
2. Climatic Design Information. Chapter 14 of ASHRAE Fundamentals Handbook 2017. ASHRAE, Atlanta, GA.
3. Huijuan Chen, D. Yogi Goswami, Elias K. Stefanakos. A review of thermodynamic cycles and working fluids for the conversion of low-grade heat. Renewable and Sustainable Energy Reviews 14 (2010) 3059–3067.
4. E.H. Wang, H.G. Zhang, B.Y. Fan, M.G. Ouyang, Y. Zhao, Q.H. Mu. Study of working fluid selection of organic Rankine cycle (ORC) for engine waste heat recovery. Energy 36 (2011); 3406 – 3418.
5. Junjiang Bao, Li Zhao. A review of working fluid and expander selections for organic Rankine cycle. Renewable and Sustainable Energy Reviews 24 (2013) 325 – 342.

Figure 3. is a schematic of an ORC cycle. The condenser rejects waste heat to the atmosphere and is a commercially available heating, ventilation, air conditioning, and refrigeration (HVACR) component. The pump increases the liquid working fluid from low to high pressure and is also commercially available. The evaporator transfers waste heat from the ICT source into ORC working fluid and is a commercially available HVACR component. The expander is the power generation component of the ORC producing electricity by driving a generator during the decrease in pressure of the working fluid vapor stream. Expanders <10kWe are not readily available commercially. [1]

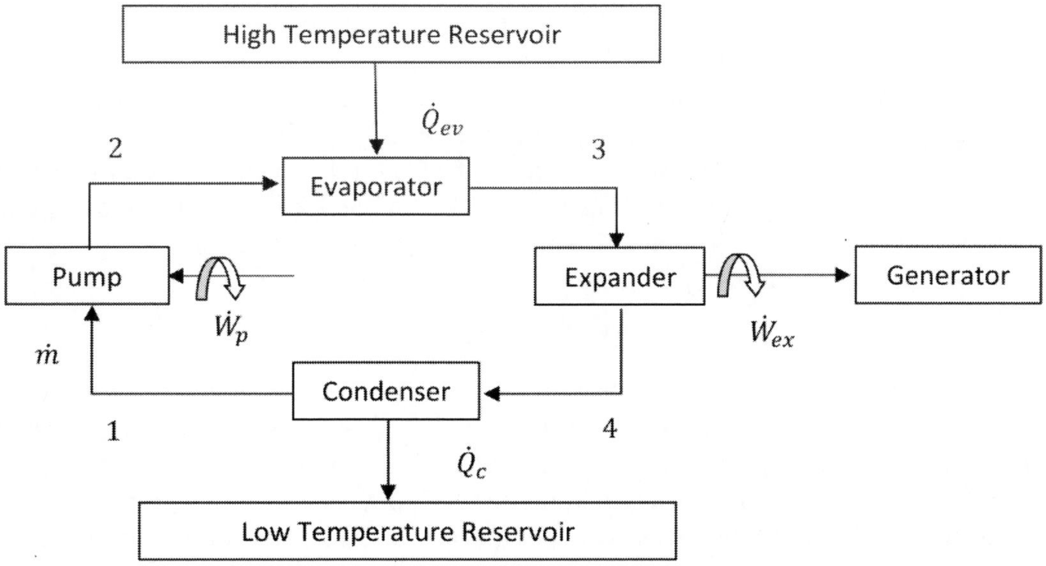

Figure 3. Schematic of ORC Heat Engine

Scroll applicability as ORC <10kWe expander

Qiu et. al. [1] paper Expanders for micro-CHP systems with organic Rankine cycle listed key findings with implications for the ICT application direction. These include:

"Micro-scale expanders, <10kWe, are not readily available on the commercial market.

Features of scroll expanders important to small-scale ORC power systems include: no valves, low parts count, low cost, and successful demonstration of efficiencies exceeding 70 percent over a wide range of rotational speeds and expansion ratios.

The working principles and characteristics of scroll expanders...make them likely choices for ORC systems within the capacity range of 1-10KWe."

1. Guoquan Qiu, Hao Liu, Sa_a Ri_at. Expanders for micro-CHP systems with organic Rankine cycle. Applied Thermal Engineering, Elsevier, 2011, 31 (16), pp.3301.

HVACR scroll compressor as expander

The expander's output power is directly proportional to the isentropic efficiency. For maximum isentropic efficiency, the expander characteristics must be matched to the operating conditions of the application. Many expanders used in research work are re-purposed compressors. Developing the expander for an ORC heat recovery system with low quality heat source is a necessary next step enabling system design and development. [1]

Few researchers design scroll expanders from scratch; rather most start with commercialized scroll compressors and operate them backwards. Two primary sources of scroll compressors are automotive and HVACR. Automotive operating life goals are measured in thousands of hours while building HVACR life goals are in tens of thousands of hours. ICT industry capital equipment is expected to operate continuously for ten years or more, 90,000 hours plus. The working fluid must be contained in this system for the entire life. A typical HVAC industry leak rate limit is <3g/year. Leak sources in order of leak quantity are: rotating seals, NPT threaded joints, SAE flare joints, and braze joints. Automotive AC uses a rotating shaft seal. Features of HVACR hermetic compressors include: means of compression and motor are enclosed in a fully sealed container, normally deep drawn steel with welded seams, only electrical feed through, and sweat suction and discharge connections. Thus, the search for acceptable scroll compressor was limited to the HVACR industry.

Methods and Materials

Down selection of ORC working fluid

Necessary properties of an ORC working fluid used in an ICT application are shown in Table 1. [2] Specific thermodynamic properties are required to achieve high operating efficiency and acceptable pressures for the ICT temperature range. Functional properties produce long operating lives through chemical stability and compatibility as well comply with EH&S requirements of zero ozone depletion potential, low global warming potential, very limited toxicity, and low flammability. [2]

Table 1. Working fluid properties necessary for ORC[1] in ICT[2] application

Thermodynamic	
a.	Isentropic or dry fluids
b.	Low liquid specific heat
c.	High latent heat
d.	Low specific volumes
e.	Moderate pressures in the heat exchangers
f.	Freezing point below the lowest operating temperature in the cycle
g.	Normal boiling point below the lowest operating temperature in the cycle
Functional	
a.	Chemical stability at the maximum operating temperature in the cycle
b.	Noncorrosive and compatible with system materials and lubricating oil
c.	Zero ozone depletion potential (ODP)
d.	Low global warming potential (GWP)
e.	Short atmospheric lifetime (ALT)
f.	Low toxicity
g.	Low, if not zero, flammability

[1] Organic Rankine cycle

[2] Information-Communications-Technologies

1. Emre Oralli. Conversion of a Scroll Compressor to an Expander for Organic Rankine Cycle: Modeling and Analysis. Thesis Submitted for the Degree of Master of Applied Science in Mechanical Engineering. University of Ontario Institute of Technology. December 2010.
2. Russell C. Tipton, Arash Golafshan, M. Toufiq Reza. Down selection of organic Rankine cycle working fluids for emerging waste heat recovery applications based on environmental, efficiency, and safety factors. Internal Report. 12-Aug-2017.

Selected working fluids for the ICT application are R1234ze, trans-1,3,3,3-tetrafluoropropene and R450A, a blend of 42% R-134a, 1,1,1,2-tetrafluoroethane / 58% R-1234ze, trans-1,3,3,3-tetrafluoropropene. [1]

Selection of expanders for test

REFPROP, REFerence fluid PROPerties, [2], and published HVACR performance data were used to evaluate efficiency of scroll compressors used as expanders. AHRI Standard 540-2015 prescribes means to calculate compressor performance. [3] ANSI/ASHRAE Standard 23.1-2010 [20] prescribes calculation of compressor efficiency. The assumption was made that, for a given volume flow rate and volume ratio, the isentropic efficiency for a scroll set would be equal whether operated as a compressor or an expander.

The volume flow rates for 50kW waste heat input using the proposed working fluids were estimated. For R450A and R1234ze(E) the volume flow ratio, $\mathring{V}$ratio, is 4.51 and 4.60, respectively. For R450A and R1234ze(E) the volume flow rate in, $\mathring{V}3$ m3/s, is 2.10 E-3 and 2.49 E-3, respectively. Assuming 55% expander isentropic efficiency, the volume flow rate out, $\mathring{V}4$ m3/s, is 9.24 E-3 and 1.12 E-2, respectively. If the expander isentropic efficiency is increased to 70%, the volume flow rate out decreases a negligible 3% on average.

Using the methodology to analyze HVACR scroll compressor performance produced a cascade of Copeland ZP91KCE-TFD and ZP182KCE-TFD. Table 2 shows the results. For R1234ze, design is 0.00249 m3/s in and 0.01120 out. ZP91 would be 0.00213 in and 0.00487 out with ZP182 0.00426 in and 0.00974 out. ZP91 is 14% undersized while ZP182 is 13% undersized. Adjusting orbit / min can accommodate this difference. The ZP91 out / ZP182 in is 1.14 which will balance to 1 during testing.

Table 2. Operation of Copeland ZP91KCE-TFD and ZP182KCE-TFD in cascade.

ZP182KCE-TFD	
$\mathring{V}_{ratio}$	2.3
Isentropic efficiency	72%
$\mathring{V}_{in}$ (m^3/s)	0.00426
$\mathring{V}_{out}$ (m^3/s), $\mathring{V}_4$	0.00974
ZP91KCE-TFD	
$\mathring{V}_{ratio}$	2.3
Isentropic efficiency	74%
$\mathring{V}_{in}$ (m^3/s), $\mathring{V}_3$	0.00213
$\mathring{V}_{out}$ (m^3/s)	0.00487

Air Squared, experimental production scroll machine manufacturer, model E25H061A-SH-CI has a volume ratio of 4.25 and $\mathring{V}3=0.0067$ m^3/s. The E25H061A-SH-CI volume flow rate is 270% and volume ratio is 92% of the ICT application design point. Reduced orbit speed will reduce the volume flow rate to ICT design.

1. Russell C. Tipton, Arash Golafshan, M. Toufiq Reza. Down selection of organic Rankine cycle working fluids for emerging waste heat recovery applications based on environmental, efficiency, and safety factors. Internal Report. 12-Aug-2017.
2. Lemmon, E.W., Bell, I.H., Huber, M.L., McLinden, M.O. NIST Standard Reference Database 23: Reference Fluid Thermodynamic and Transport Properties-REFPROP, Version 10.0, National Institute of Standards and Technology, Standard Reference Data Program, Gaithersburg, 2018.
3. AHRI Standard 540-2015 Standard for Performance Rating of Positive Displacement Refrigerant Compressors and Compressor Units. Air-Conditioning, Heating, and Refrigeration Institute. Arlington. 2015.

Testing ORC ICT application expander

As discussed, condenser, pump, and evaporator are commercially available devices while expanders are not. The expander is a technology gap and testing of the selected expanders is necessary. A gas cycle test rig can simply evaluate proposed expanders. If expander and compressor were isentropic, connecting the two would be sufficient. Neither are isentropic. Adding a heat exchanger after the compressor to eliminate the heat gain due to inefficiency of both devices will produce the necessary expander inlet conditions for testing. Figure 4, Temperature entropy diagram of expander test cycle shows this simple set up. Expander outlet enters the compressor to have pressure increased to that of expander design inlet.

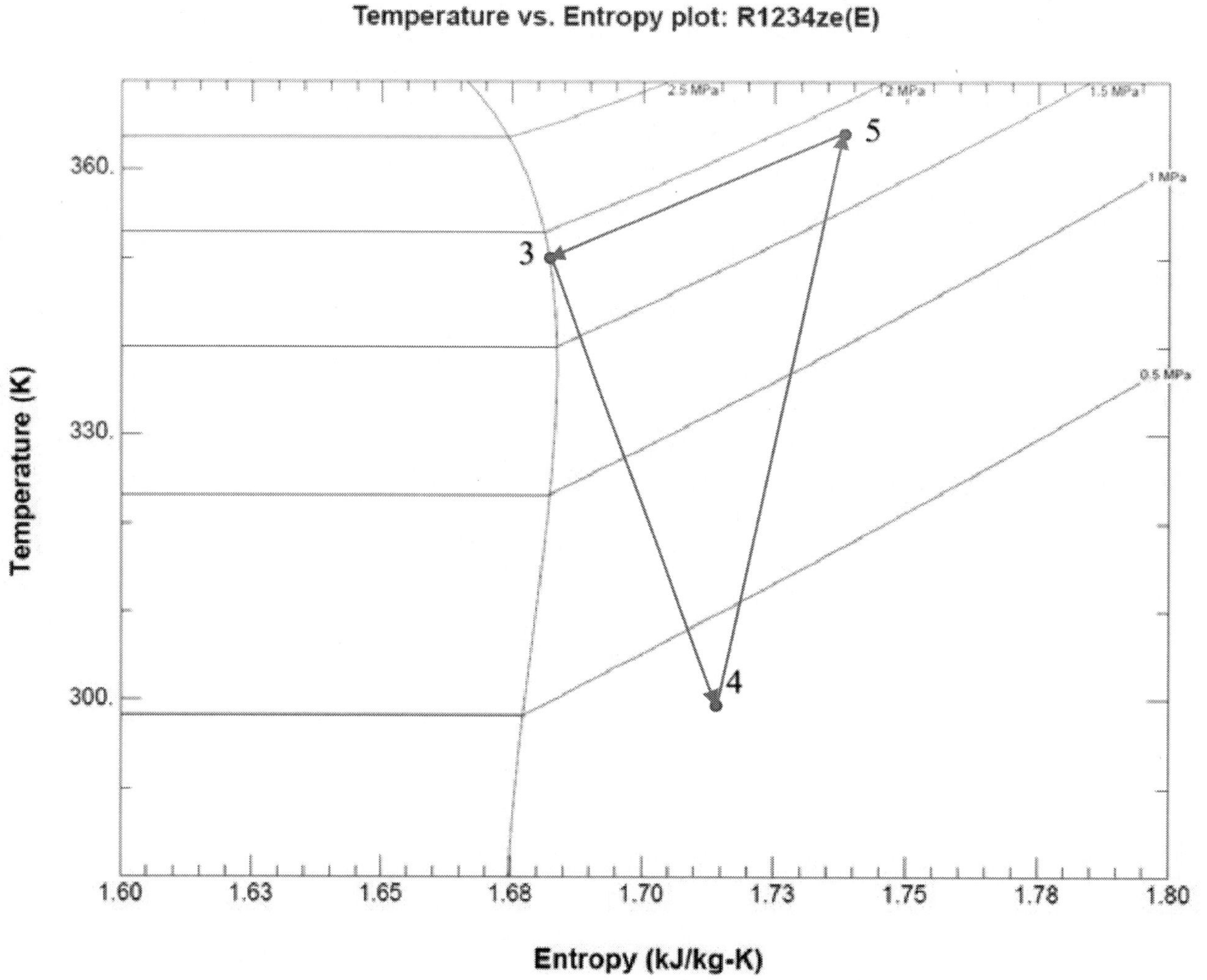

Figure 4. Temperature entropy diagram of expander test cycle

Figure 5 shows a mechanical schematic of the expander, compressor, and fan coil heat exchanger. Variable frequency drive electrical supply and rheostat expander load are also shown. Items to measure are expander inlet and outlet temperature and pressure, barometric pressure, working fluid mass flow rate, and expander electrical Volts, Amps, Watts, and Hertz output.

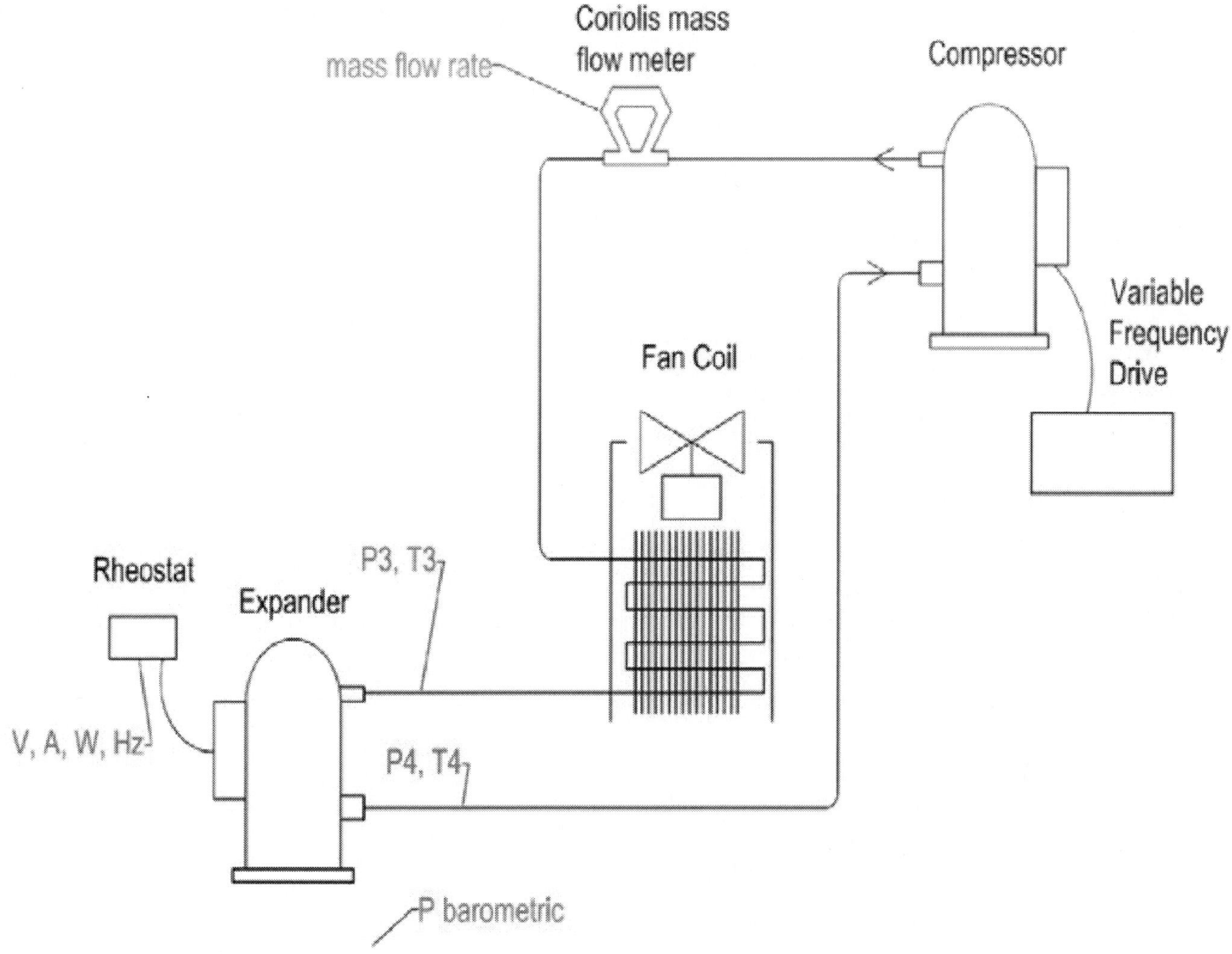

Figure 5. Gas cycle expander test schematic

Expander performance is calculated from the measured mass flow rate, pressure and temperature entering and leaving, and barometric pressure. Rheostat Volt, Amp, and Watt measurements confirm the calculated work output. The electrical frequency, Hz, indicates the scroll orbiting speed which is used to establish over- or under-expansion operation.

Discussion

An expander test matrix is shown in Table 3. An individual Copeland ZP182KCE-TFD, Copeland scroll cascade of ZP91KCE-TFD and ZP182KCE-TFD, and selected Air Squared E25H061A-SH-CI expanders will be tested. Controlling compressor speed, fan coil air flow rate, and rheostat load enables achieving entering working fluid temperature at saturation from 350K to 335K, entering volume flow rates from 90% to 110% of design, leaving flow rates 90% to 110% of design, and mass flow rates from 50% to 100% of design for both R1234ze and R450A working fluids.

Table 3. Expander test matrix.

Expanders
• Cascade Copeland ZP91KCE-TFD and ZP182KCE-TFD
• Air Squared E25H061A-SH-CI
Entering volume flow rate, m^3/s
• 0.0018 to 0.0022
Leaving volume flow rate, m^3/s
• 0.0072 to 0.011
Entering saturated vapor temperature, K
• 350, 345, 340, 335
Working fluids
• R1234ze
• R450A

Figure 6. illustrates the potential energy benefit of ORC heat recovery applied in a Data Center. For a 400kWe IT data center, the cooling energy usage is reduced from 196kWe to a net gain of 24kWe. Essentially, the heat recovery balances the lighting and power inefficiencies of the data center such that over 99% of the data center energy is applied to the IT equipment.

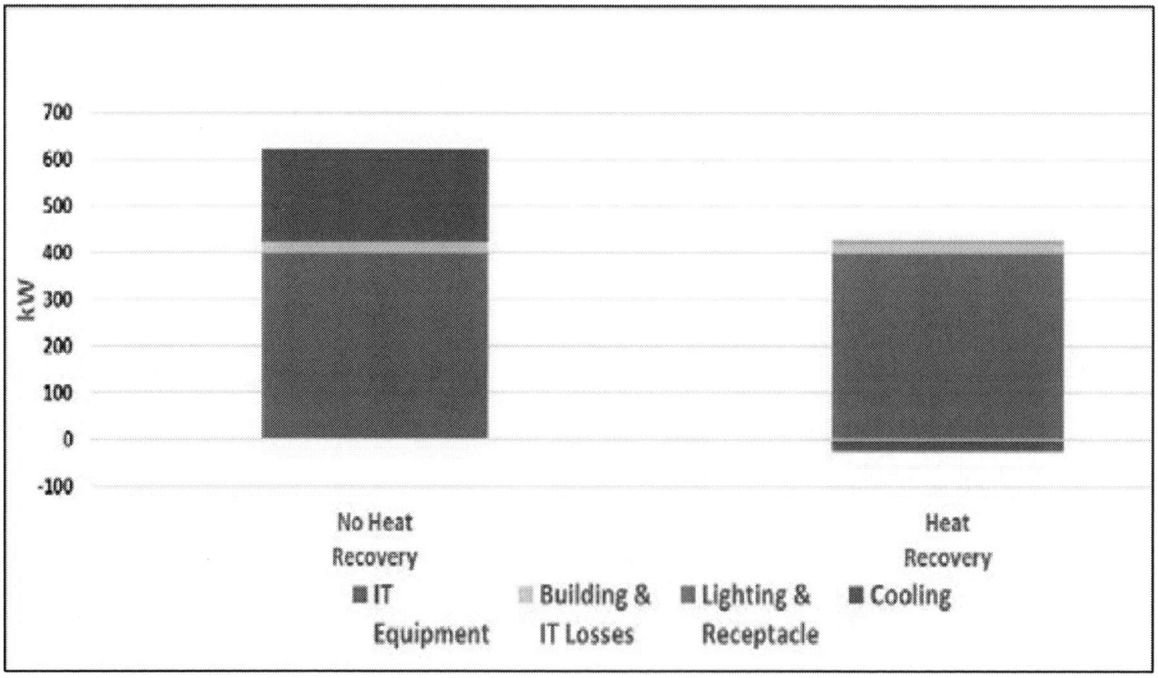

Figure 6. Data Center Heat Recovery Energy Benefit

Conclusion

- The Data Center Heat Recovery Energy Benefit is significant.
- The ORC has demonstrated capability of producing this heat recovery.
- The expander is a technology gap to commercialization of ORC. A repurposed HVACR scroll compressor is proposed for evaluation to fill this technology gap.

Future work

- Evaluation of a repurposed HVACR scroll compressors and available experimental expander will be conducted to determine their effectiveness for the application.
- The test stand of Figure 8. will be constructed.
- The expanders of Table 3. will be tested under the conditions prescribed in Table 3.

Acknowledgement

Russ Tipton acknowledges financial support for conference attendance from Vertiv.

526138 Understanding the Single Pass Operation of Vrfb and the Associated Mass Transfer Loss

Abstract:

All vanadium redox flow batteries have been recognised as an ideal energy storage device for large scale storage requirements. VRFB uses a common reservoir to store the electrolyte entering and exiting the electrodes. This results in the continuous variation of reactant concentrations and potential during a constant current operation. A constant or steady state performance can be achieved if the inlet concentration would remain constant. This is conventionally not implemented because of the larger volume of storage required and the necessity of an optimized geometry and operating parameters such that the inlet electrolyte is completely consumed within a single pass. There is also a large concentration loss associated with the low flow rates required.

The present work aims at understanding the concentration losses associated with a single pass operation with the aid of a simple lumped parameter model [1] modified to incorporate reservoir model, spatial variation, and mass transfer losses. VRFB performance optimization results based on an analysis of the integrated model will be presented. An optimum inlet concentration profile for the efficient implementation of a single pass operation will also be investigated.

References

1. Deepa Elizabeth Eapen, Suman R. Choudhury, and Raghunathan Rengaswamy. "Low grade heat recovery for power generation through electrochemical route: Vanadium Redox Flow Battery, a case study." *Applied Surface Science* (2018).

532709 CZTS (Cu_2ZnSnS_4) Electrode for Solar Rechargeable Polysulfide Bromide Redox Flow Battery

Abstract:

The increased dependency on solar energy requires large scale energy storage system as the source is often intermittent. Redox Flow Batteries (RFB) are very successful at storing the large amount of energy in a cost effective, safe and efficient environment. We took a unique approach for simultaneous solar energy conversion and storage by introducing a K doped CZTS (Cu_2ZnZnS_4: K) photo electrode in a polysulfide-bromide RFB. The photo-assisted reduction of polysulfide generating stable ~ 0.3 mA/cm^2 photocurrent corresponds to the charging of the battery. The usage of the same electrode to discharge the battery eliminated the need for an additional electrode. The catalytic effect of Cu^+ has been exploited for improving electrochemical performance of the electrode.

534772 Cuprous Bromide: An Examination of High Halide Copper Electrodeposition and Its Application in a Flow Battery

Abstract:

Recent advances in an all-copper flow battery has created a renewed interest in copper electrodeposition from high concentration (>1M) halide electrolyte[1]. This renewed interest is motivating further studies towards understanding high halide copper electrochemistry and its implications in various applications. The most apparent distinction between a high concentration halide copper electrolyte and a traditional copper sulphate electrolyte is that the high concentration halide electrolyte stabilizes the copper (I) oxidation state. There are limited reported studies in the literature of kinetics[2], diffusion[2,3], thermodynamics[4], and morphology[5] related to copper deposition from halide electrolytes. However, these studies are few and limited to the chloride chemistry. More work needs to be completed in other halides to fully understand the mechanistic nuances of these chemistries.

Since high bromide copper electrochemistry has been minimally studied, this work aims to advance our knowledge of high halide copper systems, in particular cuprous bromide. Various electroanalytical techniques were used to investigate the morphological and electrochemical differences of cuprous bromide and cupric sulphate. This presentation highlights the electrochemical and morphological differences and discusses the performance of an all-copper flow battery in a high concentration bromide electrolyte.

(1) Lloyd, D.; Magdalena, E.; Sanz, L.; Murtomäki, L.; Kontturi, K. Preparation of a Cost-Effective, Scalable and Energy Efficient All-Copper Redox Flow Battery. *J. Power Sources* **2015**, *292*, 87–94.
(2) Sanz, L.; Palma, J.; Garcia-Quismondo, E.; Anderson, M. The Effect of Chloride Ion Complexation on Reversibility and Redox Potential of the Cu(II)/Cu(I) Couple for Use in Redox Flow Batteries. *J. Power Sources* **2013**, *224*, 278–284.
(3) Ribeiro, A. C. F.; Esteso, M. A.; Lobo, V. M. M.; Valente, A. J. M.; Simões, S. M. N.; Sobral, A. J. F. N.; Burrows, H. D. Diffusion Coefficients of Copper Chloride in Aqueous Solutions at 298.15 K and 310.15 K. *J. Chem. Eng. Data* **2005**, *50*, 1986–1990.
(4) Zhao, H.; Chang, J.; Boika, A.; Bard, A. J. Electrochemistry of High Concentration Copper Chloride Complexes. *Anal. Chem.* **2013**, *85*, 7696–7703.
(5) Mackinnon, D. J.; Brannen, J. M.; McMillan, R. S. Factors Affecting the Structure of Copper Deposits Electrowon from Aqueous Chloride Electrolyte. *J. Appl. Electrochem.* **1985**, *15*, 649–658.

517109 Highly Selective Electroreduction of Carbon Dioxide into Fuels with High Current Density on Mesostructured Copper Oxide-Derived Inverse Opals

Abstract:

The production of environmentally-sustainable chemicals and fuels from CO_2 by highly efficient, selective, Earth-abundant, robust electrocatalytic materials is of vital interest not only to attenuate the climate change but also to find alternative ways to maintain carbon resources. In this work, we focus on well-defined mesostructured CuO-derived inverse opal (CuO-IO) electrode which has demonstrated remarkably high CO selectivity and negligible H_2 evolution at modest overpotentials compared to typical CuO materials. The cycle voltammetry reveals the reversible Cu^0 â†" Cu^+ â†" Cu^{2+} redox waves which is consistent with morphological and oxidation changes in CuO-IO before and after reaction observed by HR-TEM, SEM, XPS, and Raman tools. High current densities at different constant potentials are very stable for an extended period of time, reaching -10 and -27 mA cm^{-2} at -0.8 and -1.1 V *vs.* RHE. Such a porous network with voids about 200 nm in diameter surprisingly exhibits a Faradaic efficiency (FE) for CO production of 72% at -0.6 V *vs.* RHE, implying the advantage of highly ordered, interconnected mesopores in a hexagonal close packed structure in favorable production of CO and other C1-C2 products.

533852 Enhancing the Stability of High-Voltage Lithium-Ion Battery Using Sulfur-Containing Electrolyte Additives

Abstract:

Abstract. In this work, the influence of two sulfur-containing electrolyte additives, prop-1-ene-1, 3-sultone (PES) and 1, 4-bultane sultone (BS) on the cyclic stability of $LiNi_{1/3}Co_{1/3}Mn_{1/3}O_2$ (NMC111) cathode and graphite anode have been simultaneously evaluated. The capacity retention was significantly increased from 69% to 86.9% after 80 charge–discharge cycles due to the addition of 3 vol.% BS in the electrolyte, and significantly increased from 69% to 85.1% after 80 charge–discharge cycles due to the addition of 2 wt. % PES in the electrolyte. The mechanism of BS and PES improving the cyclic stability of NMC111 was investigated via theoretical calculation and experimental characterizations. The data indicates that the improved cycling stability could be ascribed to the participation of PES and BS in the formation of a thin and uniform film on the NMC111 cathode. This film effectively suppresses the subsequent decomposition of baseline electrolyte and further structure degradation of NMC111, resulting in improved cyclic stability of NMC111/Li on high work voltage. Meanwhile, according to the charge– discharge test on the graphite/Li, BS and PES-containing cells show stable cycle efficiency on graphite anode during 50 cycles.
Abstract. In this work, the influence of two sulfur-containing electrolyte additives, prop-1-ene-1, 3-sultone (PES) and 1, 4-bultane sultone (BS) on the cyclic stability of $LiNi_{1/3}Co_{1/3}Mn_{1/3}O_2$ (NMC111) cathode and graphite anode have been simultaneously evaluated. The capacity retention was significantly increased from 69% to 86.9% after 80 charge–discharge cycles due to the addition of 3 vol.% BS in the electrolyte, and significantly increased from 69% to 85.1% after 80 charge–discharge cycles due to the addition of 2 wt. % PES in the electrolyte. The mechanism of BS and PES improving the cyclic stability of NMC111 was investigated via theoretical calculation and experimental characterizations. The data indicates that the improved cycling stability could be ascribed to the participation of PES and BS in the formation of a thin and uniform film on the NMC111 cathode. This film effectively suppresses the subsequent decomposition of baseline electrolyte and further structure degradation of NMC111, resulting in improved cyclic stability of NMC111/Li on high work voltage. Meanwhile, according to the charge– discharge test on the graphite/Li, BS and PES-containing cells show stable cycle efficiency on graphite anode during 50 cycles.

521603 A Molecular Simulation Study for Natural Gas Upgrading through Mixed-Matrix Membranes Formed By PB-1A Organic Cage and a Polymer with Intrinsic Microporosity

Abstract:

In recent years, natural gas has become one of the most important sources. Compared with coal and crude oil, natural gas is environmentally more benign. Eliminating the impurities in raw natural gas is essential to meet the standard of transportation and production. Membrane-based separation has been widely used for natural gas upgrading and there is continuous quest for high-performance membranes. In this study, we computationally design mixed-matrix membranes (MMMs) from a porous organic cage (PB-1A) and a polymer of intrinsic microporosity (PIM-1), and examine their performance for natural gas upgrading by molecular simulations. The results show that the MMMs significantly improve gas permeability without sacrificing selectivity. The structural properties (e.g. free volume and pore size distribution) of the MMMs, gas sorption and diffusion are investigated to analyze the governing factors leading to the enhanced performance.

535752 Computational Screening of Hydration Reactions for Thermal Energy Storage: New Materials and Design Rules

Abstract:

Thermal energy storage (TES) materials offer a unique approach to renewable energy storage. TES materials store energy by absorbing heat when it is in excess and releasing heat when energy is needed. For instance, TES is ideal for thermal energy regulation (e.g., of buildings, textiles, and electronics), waste heat capture, and improving the efficiency of solar energy plants. Of particular interest is storing the energy in hydration/dehydration reactions since this method has the potential to store large amounts of energy in a process that is simple, cost effective, and reversible at moderate temperatures. In the current study, we performed high-throughput, first principles calculations on a set of salt hydrates and metal hydroxides to determine the most promising candidates for TES based on energy storage densities.[1] We selected known compounds from the Inorganic Crystal Structure Database, optimized their structure, and calculated their gravimetric and volumetric energy densities in addition to their operating temperature range. In total, 265 hydration reactions were characterized, and promising reactions were identified for low (<100 °C), medium (100-300 °C), and high (>300 °C) temperature ranges. Many promising reactions identified in the current study have been previously unexplored for TES applications, including the dehydration of $CrF_3 \cdot 9H_2O$. Additionally, property-performance relationships were identified using a Pearson correlation matrix. For salt hydrates, performance is highly dependent on the water capacity of the salt hydrate, and, for hydroxides, performance is strongly influenced by ionic properties of the material, such as the electronegativity of the cation. Our work represents one of the largest screening studies involving salt hydrates and metal hydroxides. The results can be used to guide experiments towards the most promising TES materials.

(1) S. Kiyabu, J.S. Lowe, A. Ahmed, and D.J. Siegel, "Computational Screening of Hydration Reactions for Thermal Energy Storage: New Materials and Design Rules," *Chemistry of Materials* 2018, 30, 2006-2017

527522 A Combined Adsorbent Bed and Pellet Model for Adsorptive Hydrogen Storage

Abstract:

A combined adsorbent bed and pellet model for adsorptive hydrogen storage

P. Sridhar and N.S. Kaisare

Department of Chemical Engineering, Indian Institute of Technology Madras, Chennai 600036, India

Micro-porous materials such as metal-organic frameworks (MOFs) and activated carbons (ACs) are considered as promising adsorbents for storage of hydrogen at cryogenic and room temperature due to their well-defined structures, high pore volumes, and high hydrogen uptake capacities. Modeling of hydrogen adsorption has been extensively studied over various scales: From molecular-level to characterize adsorption process and energetics, to entire system-level to investigate discharge and refuelling dynamics. In modelling refuelling behaviour of adsorbent bed, 2D and 3D models have been popular. Kumar (2011) investigated the hydrogen adsorption in MOF-5 flow-through bed using a 2D model. He showed that the bed requires 150 seconds to reach 5 wt% gravimetric capacity. Due to its low density and poor thermal and structural properties, MOF-5 is often pelletized and the storage bed contains numerous such pellets. Ortmann and Kaisare (2015) investigated refuelling of hydrogen in a single MOF-5 pellet, and observed that a pellet with 1.5 cm diameter took 15 seconds to achieve 5 wt% gravimetric capacity. The refuelling time increased as the pellet size increased. However, the bed-level models have often assumed local equilibrium and neglected interfacial mass and heat transfer between gas and solid particles. Furthermore, the effect of meso-scale processes, such as sorption at the adsorbent site and heat transfer and diffusion through the pellets, has been neglected in modeling and design of adsorbent beds.

Therefore, the objective of this work is to investigate multi-scale behaviour of adsorption bed by combining macroscopic device-level model with mesoscopic pellet-level model that explicitly incorporate heat and mass transfer within the pellet and sorption at the adsorption site. Thereby, we study how pellets at different location in the bed behave during adsorption of hydrogen and the effect of mass transport resistances on the system behaviour.

A mathematical model consisting of mass, momentum, and energy balance equations for hydrogen sorption tank is solved simultaneously using COMSOL Multiphysics to obtain the dynamics during refueling of the adsorbent bed. The adsorption in a spherical pellet is studied using a coupled 1D model of the spherical pellet. The macro- and micro-scale balance equations are defined in different coordinate systems: Macroscale part along the length of the bed is set up in 1D (or 2D) and the mesoscopic part along the pellet radius, thus resulting in a 1D-plus-1D (or 2D-plus-1D) model. The concentration distribution in the particle gives the molar flux at every point along the bed. At the pellet-level, the Knudsen diffusion model is applied to include mass transfer resistance in the pellet. Modified Dubinin-Astakhov adsorption isotherm model is applied to study the adsorption behavior with the variation in temperature and pressure within the pellet. The overall model is solved to investigate the effect of operating conditions, physical properties, pellet size and other parameters on refuelling dynamics, and hence optimize the hydrogen storage system.

References:

Kumar, V. Senthil. "A generalized cryo-adsorber model and 2-D refueling results." *International Journal of Hydrogen Energy* 36.23 (2011): 15239-15249.

Ortmann, Jerome P., and Niket S. Kaisare. "Modeling of cryo-adsorption of hydrogen on MOF-5 pellets: Effect of pellet properties on moderate pressure refueling." *International Journal of Hydrogen Energy* 41.1 (2016): 342-354.

526577 High Energy Density Energy Storage System Composed of Electrolyzer, Metal Hydride, and Fuel Cell

Abstract:

Hydrogen is attracting great attention as a promising energy source due to its high energy storage density, environmental friendliness, and connectivity with renewable energy sources. However, there are disadvantages in that the charging hydrogen is restricted due to the lack of infrastructure and hydrogen is difficult to handle without proper equipment. In this study, a novel type of reversible energy storage system that uses water as a reactant without hydrogen supply has been developed. This energy storage system was specialized to enhance volumetric energy storage density compared to conventional Li-ion battery in the portable application area. The core technologies of the system are water electrolyzer, metal hydride, and fuel cell. The charging unit is composed of a pressurized-type water electrolyzer and BOPs (Balance of plant) including a water booster pump and the dryer module. The discharging unit is composed of an AB_5 type metal hydride based hydrogen storage cylinder, an air-breathing type PEMFC (Polymer electrolyte membrane fuel cell) stack which can be operated without humidifying device, and a DC/DC converter. In the charging process, water is converted into hydrogen through electrochemical reaction in the electrolyzer of the charging unit and hydrogen is stored in the metal hydride of the discharging unit. During the discharge process, the stored hydrogen is converted into electricity and water through electrochemical reaction in the PEMFC stack of the discharging unit. In this study, high pressure operation of the electrolyzer, PCT (Pressure-concentration-temperature) characteristics of the AB_5 type metal hydride, and the high efficiency operating conditions of the PEMFC stack were studied. In addition, energy charging and discharging characteristics were experimentally investigated when core components were integrated. From the experimental results, effective design approach for enhancing charging/discharging characteristics of this system were proposed.

533151 Heat and Mass Transfer of Complex Metal Hydride Hydrogen Storage Reactor with Improved Heat Exchange System: Modelling and Simulation

Abstract:

Extensive research to find alternative clean and eco-friendly renewable energy sources over the years has increased interest in hydrogen as a fuel that can potentially replace traditional fossil fuels for various applications. The application of hydrogen fuel technology in the automotive industry is limited by challenges arising from the storage of hydrogen. Solid-state hydrogen storage is one possible solution, where a porous metal hydride material is packed into a bed in a reactor; however, current hydrogen storage systems based on this method still exhibit high charge/discharge temperature ranges and slow sorption kinetics. This thermodynamic and kinetic limitations of metal hydride storage methods leave a room for further research into finding better storage materials in combination with optimal reactor designs. Hydrogen charge/discharge from the metal hydride bed depend on the macro-kinetics and physical adsorption/desorption processes which are related to the thermal management of the storage system

In this work, a hydrogen storage reactor with a combination of internal helical and external jacket heat exchange configurations packed with a Li-Ni-Mg based storage material was tested both experimentally and theoretically. Heat and mass transfer simulation results of this system (via OpenFOAM) were in agreement with our experimental data as well as literature findings. The combination of CFD and experiment were used to develop an improved reactor thermal management and material sorption kinetics.

533828 A Symmetrical Solid Oxide Fuel Cell with a-Site Sodium Doped Perovskite Electrode Materials

Abstract:

Solid oxide fuel cells (SOFCs) are electrochemical energy conversion devices those directly convert chemical energy in fuels to electrical energy with a high efficiency and low emissions. Recently, symmetrical SOFCs (SSOFCs) with a simplified configuration using the same material as both cathode and anode have raised much attention. The biggest challenge to the electrode material for a SSOFC, however, is that it must maintain structural stability and electrocatalytic activity when operating in both reducing and oxidizing environment. In this work, A-site sodium doped double-perovskites $Sr_{2-x}Na_xFeMo_{0.65}Ni_{0.35}O_{6-d}$ (SN_xFMNi, x=0.05, 0.1 and 0.2) are synthesized and investigated as electrode materials for SSOFCs. All the samples undergo the same phase transition process from a perovskite structure to a Ruddlesden-Ropper phase (RP phase) with the exsolution of numerous Ni_3Fe nanoparticles during 10 h reduction. The phase transition process is studied with X-ray diffraction, thermogravimetric analyzer and transmission electron microscope, and the results indicate that all the samples keep a perovskite structure in an oxidizing environment while turn into a RP phase in a reducing environment. Amongst all the samples, $SN_{0.2}FMNi$ shows the lowest activation energy of 0.30 eV. An electrolyte-supported symmetrical single cell with a configuration of $SN_{0.2}FMNi|LDC|LSGM|LDC|$ $SN_{0.2}FMNi$ gives the maximum power densities of 1157 and 335 mW cm^{-2} with H_2 and humidified CH_4 as the fuels at 850 °C, respectively.

529935 Humidity Tracking By Mixing Dry and Humidified Gases with Internal Model Control for PEM Fuel Cells

Abstract:

Humidity tracking by mixing dry and humidified gases with internal model control for PEM fuel cells

Sathish Swaminathan [a], Srinivasan Raman [b], Raghunathan Rengaswamy [a,b] *

[a] Department of Chemical Engineering, Indian Institute of Technology â€“ Madras, Chennai, India

[b] Department of Chemical Engineering, Texas Tech University, Lubbock, Texas

* raghur@iitm.ac.in, raghu.rengasamy@ttu.edu

Abstract

Humidification of gases is widely employed in various manufacturing industries, medical and energy equipment including proton exchange membrane (PEM) fuel cells. Humidification of fuel cell reactants is essential for maintaining the proton conductivity. PEM fuel cells produce excessive water under high current operation which sometimes result in detrimental operating conditions. Hence, controlling the humidity in reactants gases is essential to maintain optimal functioning of the fuel cell system. Changes to humidity of gases can be achieved by mixing of dry and humidified gases as shown in literature. The reactants passing through the humidifier for absorbing water vapors, induces a time delay in the humidified gases before reaching the fuel cells. It was experimentally observed that this time delay due to humidification results in a mismatch between the humidity set points and output, while using a conventional PI/PID controller for tracking humidity changes.

In this work, the humidity of a stream of gases was controlled using internal model control to eliminate the time delays in set point tracking, before supplying the gases to a fuel cell system. Experiments were conducted to obtain a desired humidification level in the gases by mixing dry and humidified gases, as decided by the controller. Accurate tracking of step and sinusoidal changes in humidity set points were achieved. The advantages, challenges, and limitations of using internal model control strategy for humidity set point tracking will be highlighted.

541565 25 by 25: Chemical Engineering in the Next 25 Years

Abstract:

In 2008, the centennial year for AIChE, we asked 25 chemical engineers for their visions of the profession from around the US and the world: Nine faculty, eight from industry, and eight post-docs and graduate students. We posed four questions related to each individual's industrial sector or area of research, asking for extrapolation, new impacts on existing sectors, new sectors, and an optional, more open-ended comment on the future of the profession as a whole.

In this 110th year of AIChE, we have revisited these questions with many of the same people. Ten years has seen many developments in our profession, in society, and in our thought-leaders' experiences. Here we report on their perspectives, comparing them to the 2008 visions.

541542 The Future of Chemical Engineering Itself

Abstract:

Chemical engineeringâ€™s future is bright because it is vital to so much of society and industry -- and its future is complicated by its rapidly expanding universe of activities. It must represent depth in core skills centered on the chemical sciences including biology and materials. At the same time, it must be transdisciplinary; it must exemplify the need and value of reaching across disciplines in borderless ways.

One way is to create and disseminate a compelling vision and strategy for chemical engineering, research, and education for the future. AIChE and the National Academies are working together to that end. In 1988, the National Academies published "Frontiers in Chemical Engineering: Research Needs and Opportunities.â€ Its committee of authors outlined a wide-ranging roadmap to turn promising research opportunities into reality, while calling for researchers to embrace new frontiers.

Since 1988, that has happened -- along with many other transformative changes like the Web, fracking, personalized medicine, and social media. A new nationwide effort, aided by international contributions, will address both the identity of chemical engineering and its future directions. Its goal is to ensure that the people in chemical engineering are primed to meet society's needs and rapid changes.

544708 Accelerating Innovation through Academic-Industrial Partnerships

Abstract:

In 2011, The Dow Chemical Company launched the University Partnership Initiative (UPI) to sponsor collaborative research at leading U.S. Universities. This 10-year commitment aims to address fundamental questions in industrial chemistry and chemical engineering with alignment to Dowâ€™s next-generation process technology and product development efforts. To date, this portfolio of projects has generated an impressive array of intellectual property, publications, and numerous examples of the collaborative developments implemented at Dow. This presentation will introduce UPI program and highlight several contributions from our collaborators and leaders in the chemical engineering profession.

540559 Maximizing Uptime, Efficiency, and Safety of Industrial Operations through Early Risk Detection

Abstract:

Achieving zero incidents while maximizing process uptime and efficiency is a key vision for operating companies. An important building block towards this vision is establishment of a proactive risk mitigation culture, supported by effective systems and a strong management. Studies have shown that most of the incidents and unexpected process failures can be avoided if the operating teams get timely information about developing risks and take preventive actions early on. While advances have been made in process monitoring and automation in the past decades, there still remains significant technological and behavioral gaps that prevent plant operations from proactively mitigating process risks.

A modern industrial plant monitors thousands of parameters, generating upwards of 50-100 million data points every day. Thanks to recent breakthroughs in machine learning and artificial intelligence approaches, today there are autonomous systems that can sift through this data and point out meaningful and timely insight. This can help operating teams ascertain process issues that are hidden in the data, long before process variables reach critical levels.

In this presentation, a new autonomous system, Dynamic Risk Analyzer™ (DRA), will be introduced that points out early indicators of process issues using proprietary machine learning. The risk indicators enable operations team to respond and make required process changes, days and sometimes weeks ahead of any alarm, avoiding late stage (expensive) fixes. To realize these objectives fully, a proactive management workflow must also be in place. The presentation will discuss new initiatives and workflows developed by plant operations to implement a proactive culture. Real-life case studies from plants using DRA technology will be presented on how this culture change resulted in increased uptime, efficiency and safety of plant operations.

531309 Gaussian Processes for Hybridizing Analytical & Data-Driven Decision-Making

Abstract:

Surrogate models are widely appreciated in process systems engineering [1]. The typical setting focuses on expensive-to-evaluate, possibly uncertain functions. Examples include: modular process simulators [2], integrated gasification combined cycle processes [3], a carbon capture absorber [4], and many other applications, e.g. [5-8]. Resources are typically limited, so effective decision making requires data-efficient learning.

The data science and statistical machine learning communities typically focus on models learned solely from observed data. But chemical engineering applications may also require explicit, parametric models, e.g. modeling known process constraints, operations constraints, and cost objectives [9]. So, work has integrated semi-algebraic functions with those learned from data [10] or developed semi-physical modeling techniques [11, 12].

This presentation surveys the state-of-the-art in hybridizing analytical and data-driven decision making [13]. We consider three probabilistic modeling applications to these hybrid situations:

Design of experiments for model discrimination [14]. We bridge the gap between classical, analytical methods [15] and Monte Carlo-based approaches [16]. Classical methods may have difficulty managing non-analytical model functions and data-driven Monte Carlo approaches come at a high computational cost. We replace the original, parametric models with probabilistic, non-parametric Gaussian process surrogates learned from model evaluations. The surrogates are flexible regression tools that extend classical analytical results to non-analytical models, while providing us with model prediction confidence bounds and avoiding the computational complexity of Monte-Carlo approaches.

Multi-objective optimization [17, 18]. We make novel extensions to Bayesian multi-objective optimization in the case of one analytical objective function and one black-box, i.e. simulation-based, objective function. The resulting method has been applied to a bone neotissue application [19] and a more general test suite.

Scheduling plant operations under uncertainty. For processes with equipment degradation, we use Gaussian processes to approximate large-scale, mixed-integer optimization problems.

We close by offering a broad outlook on applying probabilistic surrogate models to chemical engineering. Statistical machine learning has recently attracted significant interest in process systems engineering [20]. Here we show that state-of-the-art research in Gaussian processes [21, 22] and probabilistic modeling more generally [23] can have a big impact on chemical engineering.

References

[1] Bhosekar A, Ierapetritou M. Advances in surrogate based modeling, feasibility analysis and and optimization: A review. *Comput Chem Eng*. 108: 250-267, **2018**.

[2] Caballero JA, Grossmann IE. An algorithm for the use of surrogate models in modular flowsheet optimization. *AIChE J*. 54(10):2633-2650, **2008**.

[3] Lang Y, Zitney SE, Biegler LT. Optimization of IGCC processes with reduced order CFD models. *Comput Chem Eng*. 35(9):1705-1717, **2011**.

[4] Cozad A, Sahinidis NV, Miller DC. Learning surrogate models for simulation-based optim-ization. *AIChE J*. 60(6):2211-2227, **2017**.

[5] Boukouvala F, Ierapetritou M. Surrogate-based optimization of expensive flowsheet modeling for continuous pharmaceutical manufacturing. *J Pharm Innov*. 8: 131, **2013**.

[6] Soepyan FB, Cremaschi S, Sarica C, and others. Estimation of percentiles using the Kriging method for uncertainty propagation. *Comput Chem Eng*. 93:143-59, **2016**.

[7] Shokry A, Ardakani MH, Escudero G, Graells M, Espuña A. Dynamic Kriging-based fault detection and diagnosis approach for nonlinear noisy dynamic processes. *Comput Chem Eng*. 106:758-76, **2017**.

[8] Tran AP, Georgakis C. On the estimation of high-dimensional surrogate models of steady-state of plant-wide processes characteristics. *Comput Chem Eng*. DOI 10.1016/j.comp-chemeng.2018.02.014, **2018**.

[9] Boukouvala F, Hasan MF, Floudas CA. Global optimization of general constrained grey-box models: new method and its application to constrained PDEs for pressure swing adsorption. *J Glob Optim*. 67(1-2): 3-42, **2017**.

[10] Boukouvala F, Floudas CA. ARGONAUT: AlgoRithms for Global Optimization of coNstrAined grey-box compUTational problems. *Optim Lett*. 11(5): 895-913, **2017**.

[11] Pearson RK, Pottmann M. Gray-box identification of block-oriented nonlinear models. *J Process Contr*. 10(4):301-315, **2000**.

[12] Cozad A, Sahinidis NV, Miller DC. A combined first-principles and data-driven approach to model building. *Comput Chem Eng*. 73: 116-27, **2015**.

[13] Boukouvala F, Misener R, Floudas CA. Global optimization advances in mixed-integer nonlinear programming, MINLP, and constrained derivative-free optimization, CDFO. *Eur J Oper Res*. 252: 701-727, **2016**.

[14] Olofsson S, Deisenroth MP, Misener R. Design of Experiments for Model Discrimination Hybridising Analytical and Data-Driven Approaches. *arXiv:1802.04170*. **2018**.

[15] Buzzi-Ferraris G, Forzatti P, Emig G, Hofmann H. Sequential experimental design for model discrimination in the case of multiple responses. *Chem Eng Sci*. 39(1): 81-85, **1984**.

[16] Vanlier J, Tiemann CA, Hilbers PAJ, van Riel NAW. Optimal experiment design for model selection in biochemical networks. *BMC Syst Biol*, 8(20), **2014**.

[17] Olofsson S, Mehrian M, Geris L, Calandra R, Deisenroth MP, Misener R. Bayesian Multi-Objective Optimisation of Neotissue Growth in a Perfusion Bioreactor Set-Up. *Computer Aided Chemical Engineering*. 40: 2155-2160, **2017**.

[18] Beykal B, Boukouvala F, Floudas CA, Pistikopoulos EN. Optimal design of energy systems using constrained grey-box multi-objective optimization. *Comput Chem Eng*. DOI 10.1016/j.compchemeng.2018.02.017, **2018**.

[19] Mehrian M, Guyot Y, Papantoniou I, Olofsson S, Sonnaert M, Misener R, Geris L. Maximizing neotissue growth

kinetics in a perfusion bioreactor: An *in silico* strategy using model reduction and Bayesian optimization. *Biotechnol Bioeng*. 115(3):617-29, **2018**.

[20] Lee JH, Shin J, Realff MJ. Machine learning: Overview of the recent progresses and implications for the process systems engineering field. *Comput Chem Eng*. DOI 10.1016/j.compchemeng.2017.10.008, **2017**.

[21] Damianou A, Lawrence N. Deep Gaussian processes. In *Artificial Intelligence and Statistics*. 207-215, **2013**.

[22] Deisenroth MP, Ng JW. Distributed Gaussian processes. In *32nd International Conference on International Conference on Machine Learning*. 37: 1481-1490, **2015**.

[23] Ghahramani Z. Bayesian non-parametrics and the probabilistic approach to modelling. *Phil Trans R Soc A*. 371(1984):20110553, **2013**.

540079 Augmentation of the Rate of Mass transfer Limited Liquidâ€“Solid Heterogeneous Reactions by Turbulence Promoters

Abstract:

Intensifications of the rate of mass transfer limited electrochemical and catalytic reactions taking place at the external surface of a helical coil imbedded in an inert fixed bed of cylinders acting as turbulence promotors was studied by the electrochemical technique where the rate of mass transfer was obtained from the limiting current of the cathodic reduction of $K_3Fe\ (CN)_6$ in presence of NaOH as a supporting electrolyte.

Variables studied were velocity of the solution, packing cylinder diameter, helical coil tube diameter and the solution physical properties. Results revealed that the mass transfer coefficient at the outer surface of the helical coil increases with increasing solution velocity and decreasing packing cylinder diameter, helical coil tube diameter was found to have a slight effect on the mass transfer rate. The following dimensionless equation was found to fit the data

$$Sh = 7.14*10^{-4} * Sc^{0.33} * Re^{0.52} * (d_p/d)^{-2.32}$$

The importance of this equation in the scale up and operation of the present reactor was pointed out. The possible use of the coil inner surface as a builtin heat transfer facility for conducting exothermic electrochemical and catalytic reactions which need rapid cooling was highlighted. The possibility of using multi imbedded coaxial helical coils of different coil diameter in practical catalytic reactors to increase their rate of production was noted.

527207 Modeling and Experiments on the Effects of Increasing Flow Baffles on Dead Zone and Growth Performance of Microalgae in a Raceway Photobioreactor

Abstract:

Microalgae for biodiesel production can be grown in an open pond or closed system photobioreactors. Closed system photobioreactors (PBR) are expensive to build and operate, while open system photobioreactors are less expensive to construct per unit volume and directly take advantage of the available sunlight and carbon dioxide (CO2) in the atmosphere. The most common open pond system according to the literature is the raceway photobioreactor (RWP). But limitations of the RWP include poor mixing and dead zones, which are areas with velocity of less than 0.1 m/s. This can lead to low biomass productivity, as compared with the closed system photobioreactor. Common improvements to the RWP to counteract dead zones involve the use of more than one paddle wheel, and introduction of sloping baffles and flow deflectors.

This research work focuses on flow optimization using Computational Fluid Dynamics (CFD) COMSOL Multiphysics to reduce dead zones and improve productivity of the reactor. The effects of variable number of flow baffles, central wall thickness, first channel diameter (distance between the first baffle and central wall), and ratio of first channel diameter to central wall thickness (RÂin/Tw) on the percentage of dead zone volume were studied. The results show that, compared to control RWP (RWP with no baffles), dead zone in the RWP modified with four baffles was decreased by 96% while the optimum ratio of RÂin/Tw was found to be between 3 and 4.8. To validate the simulation results, experimental results in a pilot â€“scale 200 liter RWP showed 50% and 40% increase in the productivity and specific growth rate of the microalgae respectively in an RWP with three flow baffles as compared with the control RWP.

534367 Implementation of a Flux-Dependent Anisotropic Diffusivity Model into Resolved-Particle CFD

Abstract:

Implementation of a flux-dependent anisotropic diffusivity model into resolved-particle CFD

Behnam Partopour, Anthony G. Dixon

Department of Chemical Engineering, Worcester Polytechnic Institute, Worcester, MA

Species diffusion inside catalyst particles is often modeled using an effective diffusivity, which can be derived from the Dusty Gas Model. Conventional assumptions in these models are 1- the diffusion is isotropic, 2- the ratio of species fluxes in the computational domain is fixed and equal to their stoichiometric ratio. In this study, we examine the accuracy of these assumptions, specifically for 3-dimensional resolved-particle CFD models. A new approach is developed to calculate a vector of fluxes for each species in each computational cell. Then the diffusivity tensor components are calculated and passed to the solver. The iterative approach continues until the diffusivity coefficients in each computational cell become constant. To make the model computationally feasible an appropriate transformation is used to change the anisotropic (full) diffusivity tensor to orthotropic (diagonal) tensor.

The method is implemented for resolved-particle CFD simulations of ethylene oxidation in small packed beds of spherical and cylindrical particles. The results show that the effects of the flux changes in the particles are considerable and reaction, species and temperature profiles are affected.

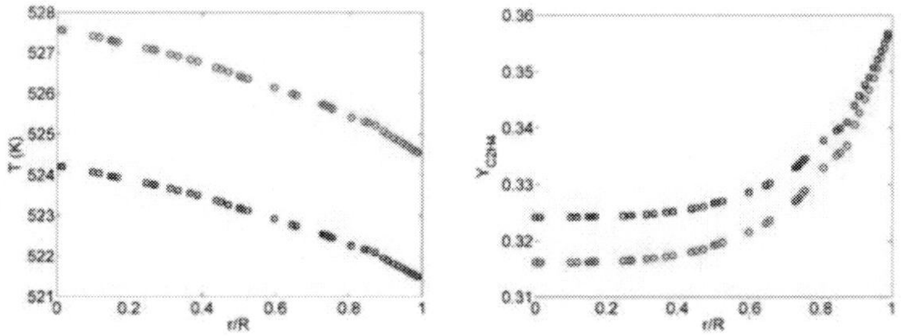

Figure. Temperature and ethylene profiles on an arbitrary radius of a selected spherical particle. Blue: flux dependent anisotropic model; Red: fixed flux isotropic model.

536786 Experimental and Numerical Heat Transfer Investigation in a Mixing Vessel with Cooling Jacket

Abstract:

Experimental and Numerical Heat Transfer Investigation in a Mixing Vessel with Cooling Jacket

Thomas Eppinger, Alexander Heyter, Ravindra Aglave, Stefan Wollny

Although agitated mixing vessels are widely used in the chemical, pharmaceutical and food industry, their design with respect to adding or removing heat is still connected to uncertainties and a subject of research. For example the standard work for engineering heat transfer problems, VDI heat atlas, lists different correlation for the heat transfer coefficient, which differ for a given Re-number by a factor of approximately 3. The reason for this are certain homogeneity assumption which do not hold.

In this contribution we are investigating the heat transfer in a heated mixing vessel with a Rushton turbine and a water cooled cooling jacket. 3D CFD simulations are validated with respect to power input, heat transfer and residence time distribution in the cooling jacket with detailed experimental data from the lab. Based on the detailed information from the CFD simulation the local heat transfer coefficient on the inner as well as on the outer side wall is analyzed and the non-homogeneities can be quantified. As a side effect, also the appropriate CFD cell resolution to capture power input and heat transfer is discussed.

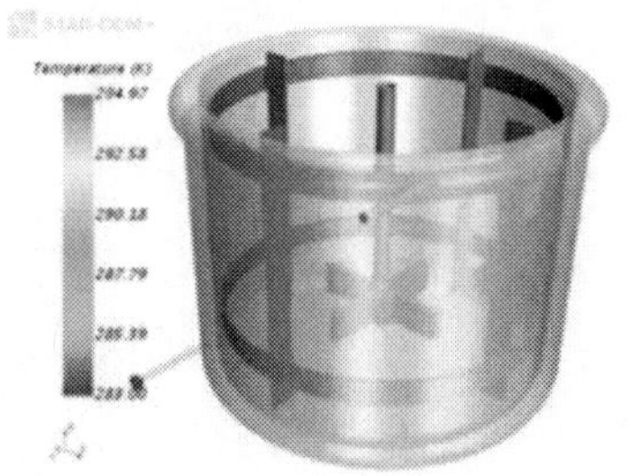

Figure 1: Fully-baffled mixing vessel with cooling jacket

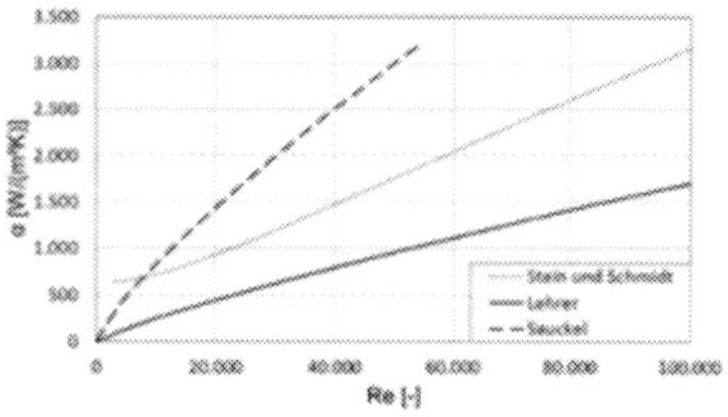

Figure 2: Heat transfer coefficient as function of impeller Re-number according to VDI heat atlas

538922 Integration Options of Electrochemical Hydrogen Pumping

Abstract:

Hydrogen is a key component in the chemical, petrochemical and biomass industry, in processes such as hydrocracking, catalytic reforming and hydrodeoxygenation. A variety of renewable and fossil-based hydrogen production technologies exist with the costs mainly depending on the hydrogen source [1]. Pressurized hydrogen production costs make up for a considerable portion of the overall operation expenditures [2] and therefore, increasing hydrogen pressurization and recirculation effectiveness is crucial for process economics. An electrochemical hydrogen compressor (EHC) is a promising technology that combines hydrogen separation and purification within a compact piece of equipment. It is a device similar to a proton exchange membrane fuel cell (PEMFC) and involves no moving parts. A membrane acts as an electrolyte for proton exchange and is positioned between two catalyst containing electrodes (usually with Pt as catalyst [3]), which in turn are positioned between two porous gas diffusion layers [4]. The electrochemical compression is an one-stage, isothermal process that requires about half of the adiabatic power needed for fixed compression ratio, at constant temperature [5].

During the last years, multiple projects have been funded, aiming at the commercialization of this application. In the USA, Analytic Power Corp. [6] and Fuel Cell Energy [7] are some of the corporations involved in such projects. In Europe, Hydrogen Efficiency Technologies (HyET, the Netherlands) demonstrated hydrogen electrochemical compression up to 1000 bar using a single-stage compressor [8]. Lower power requirements, simplicity of the system and hydrogen purification are the major advantages that triggered research into this application.

Electrochemical hydrogen compressors are mainly classified based on the operating temperature: low-temperature EHC refers to electrochemical pumping conducted at 50 - 80 °C using Nafion® membranes as electrolytes, whereas high-temperature EHC (conducted at >100 °C) uses polybenzimidazole (PBI) membranes. Several losses and irreversibilities have to be taken into account, irrespective of the operating temperature before integrating an EHC into existing process flowsheets. Losses associated with the proton resistance of the membrane (Ohmic losses), as well as activation losses that refer to the energy associated with the activation of the anode and cathode reactions, have to be taken into consideration. In addition, due to the anode and cathode pressure difference, molecular hydrogen flows back from the cathode to the anode department, which results in lower actual compressed hydrogen. The selection of the temperature range presents several characteristics and operation modes, which are different at each EHC type. For example, operating at low temperatures render the anode catalyst susceptible to CO and other impurities, whereas at higher temperatures the catalyst tolerance increases drastically. In addition, Nafion® membranes require hydration in order to reduce the losses associated with proton conductivity across the membrane, which is mainly conducted by humidifying the inlet gaseous mixture. This hydration requirement needs proper management in order to prevent drying of the anode or liquid water accumulation at the cathode department. On the other hand, high-temperature operation requires no hydration, whereas at normal conditions, water exists only in gaseous phase. However, higher outlet pressures are reported for low-temperature EHC, making them suitable for high pressure refinery and biorefinery applications. Figures 1-4 present some example results of operation characteristics of the low-temperature EHC operation.

In this work, EHC process integration options are explored, based on process modelling accounting for all important phenomena and losses in such systems, including water management for the low-temperature EHC. The proposed model enables the investigation of a broad range of operating parameters and conditions, while it can also be easily integrated in flowsheets representing wider process systems. In addition, case studies are conducted that investigate the combination of electrochemically compressing (either low- or high-temperature) up to a certain intermediate pressure and then mechanically compressing H_2 to the required, higher pressure.

References

[1] P. Nikolaidis and A. Poullikkas, 'A comparative overview of hydrogen production processes', *Renewable and Sustainable Energy Reviews*, vol. 67. pp. 597–611, 2017.

[2] Y. Zhang, T. R. Brown, G. Hu, and R. C. Brown, 'Techno-economic analysis of two bio-oil upgrading pathways', *Chem. Eng. J.*, vol. 225, pp. 895–904, 2013.

[3] A. Tokarev and D. G. Bessarabov, 'Modeling of bimetallic Pt-based electrocatalyst on extended-surface support for advanced hydrogen compression and separation', *Int. J. Hydrogen Energy*, vol. 39, no. 15, pp. 7805–7810, 2014.

[4] S. J. Hamrock and M. a. Yandrasits, 'Proton Exchange Membranes for Fuel Cell Applications', *J. Macromol. Sci. Part C*, vol. 46, no. 3, pp. 219–244, 2006.

[5] K. Onda, K. Ichihara, M. Nagahama, Y. Minamoto, and T. Araki, 'Separation and compression characteristics of hydrogen by use of proton exchange membrane', *J. Power Sources*, vol. 164, no. 1, pp. 1–8, 2007.

[6] B. S. Mackenzie and D. P. Bloomfield, 'Electrochemical hydrogen compressor. DOE Grant DE-FG02-05ER84220 to Analytic Power Corp. Final Report.', 2006.

[7] L. Lipp, P. Patel, E. Sutherland, and D. Peterson, 'Electrochemical Hydrogen Compressor - DOE Hydrogen and Fuel Cells Program FY 2013 Annual Progress Report'.

[8] P. Bouwman, 'Electrochemical Hydrogen Compression (EHC) solutions for hydrogen infrastructure', *Fuel Cells Bull.*, vol. 2014, no. 5, pp. 12–16, 2014.

536781 Advanced Water Management in Polymer Electrolyte Fuel Cells Using Engineered Gas Diffusion Layers with Patterned Wettability

Abstract:

The water management in polymer electrolyte fuel cells (PEFCs) is a rather complex issue [1]. On one hand side, some water is needed to hydrate a proton conductive membrane and ionomer in the catalyst layers. Therefore, humidified gases are fed into the cell. On the other hand, excessive amounts of water in the porous layers lead to blockage of available pores for gas transport and consequently increased mass transfer limitations, especially at high current densities when large amounts of water are produced. The gas diffusion layers (GDLs) are highly porous carbon fiber materials which are sandwiched between electrodes and flow fields and provide several functionalities: conduction of electricity and heat, mechanical integrity of the membrane, distribution of reactant gases and removal of the liquid water produced. These functions imply a triple set of contradictory requirements: high diffusivity of the gas phase, high permeability of the water phase, and high thermal and electrical conductivity of the solid phase [2].

The creation of artificial pathways for water removal within the porous material can enable the development of advanced water management strategies; however none of the existing approaches proposed a method to produce such materials in a way compatible with mass productions. To tackle this issue, we invented a method to produce GDLs with patterned wettability based on localized electron radiation grafting [2]. The method permits creating dedicated through-volume hydrophilic patterns for selectively removing water, while the remaining material maintains its hydrophobic character. In this talk, the details of the synthetic method will be discussed first [3]. Later, electrochemical characterization of *operando* cells combined with neutron radiography will be discussed to demonstrate the potential of these materials to improve power density at various conditions due a reduced mass transfer overpotential [4].

References

[1] T. V. Nguyen et al., **1993** *J. Electrochem. Soc. 140(8), 2178-2186.*

[2] A. Forner-Cuenca et al., *Adv. Mater.* **2015**, 27, 6317-6322.

[3] A. Forner-Cuenca et al., *J. Electrochem. Soc*. **2016**, 163 (8), F788-F801.

[4] A. Forner-Cuenca et al., *J. Electrochem. Soc*. **2016**, 163 (13), F1389-F1398.

537733 A Low-Dimensional Electrochemical Model for Scaling and System Analysis of Redox Flow Batteries

Abstract:

Redox flow batteries (RFBs) are promising electrochemical devices for grid-scale energy storage due to independent scaling of power and energy, capacity retention, and simplified manufacturing, but further cost reductions are needed for widespread deployment [1]. While many research efforts have focused on identifying new redox couples or increasing power output of single flow cells, less effort has been directed toward translating data obtained in laboratory-scale cells to at-scale electrochemical stacks as this is typically reserved for those commercializing the technology [2]. However, developing these rigorous scaling relationships is a critical step to predicting stack performance based on single cell data, which, in turn, enables early-stage decision-making before significant time or capital investment. Since commercial deployment has been limited, experimental data on stack performance is typically unavailable, motivating the development of modeling frameworks to bridge the gap from single cell to stack. Unfortunately, current multi-cell models are often overly simplified to the point where critical processes are not accurately described; whereas, single cell models are typically multidimensional which challenges scalability and do not capture several stack-specific issues (e.g., non-uniform flow distributions, shunt currents, and manifold design) [3,4]. Indeed, robust low-dimensional models, which account for relevant material properties and processes, may enable the accurate prediction of system-level performance, from laboratory inputs, hence accelerating RFB development.

This presentation will describe the development and validation of a scalable, low-dimensional, physics-based model that enables the translation of single cell experiments to low-error estimates of stack performance and suitable guidelines for RFB design and operation. Experimental measurements of the physical and electrochemical properties of redox electrolytes and small-volume cell performance are combined with a one-dimensional electrochemical model to quantify relevant dimensionless relationships and scaling factors [4]. These results are then applied to a scaled stack geometry to evaluate system-level performance metrics as functions of material properties and operating parameters. Integration of additional system losses, such as shunt currents and crossover rates, yields accurate descriptions of system behavior at low levels of complexity. Finally, examination of the resulting stack performance enables generalized scaling relationships and heuristics to guide electrochemical reactor design and operation.

References

[1] A.Z. Weber, M.M. Mench, J.P. Meyers, P.N. Ross, J.T. Gostick, Q. Liu, Redox flow batteries : a review, J. Appl. Electrochem. 41 (2011) 1137–1164. doi:10.1007/s10800-011-0348-2.

[2] J. Winsberg, T. Hagemann, T. Janoschka, M.D. Hager, U.S. Schubert, Redox-Flow Batteries: From Metals to Organic Redox-Active Materials, Angew. Chemie - Int. Ed. 56 (2017) 686–711.

doi:10.1002/anie.201604925.

[3] R.M. Darling, H.-S. Shiau, A.Z. Weber, M.L. Perry, The Relationship between Shunt Currents and Edge Corrosion in Flow Batteries, J. Electrochem. Soc. 164 (2017) E3081–E3091. doi:10.1149/2.0081711jes.

[4] J.D. Milshtein, K.M. Tenny, J.L. Barton, J. Drake, R.M. Darling, F.R. Brushett, Quantifying Mass Transfer Rates in Redox Flow Batteries, J. Electrochem. Soc. 164 (2017) 3265–3275. doi:10.1149/2.0201711jes.

547350 Organizing Platinum Metal and Ionomer with Engineered Protein

Abstract:

We engineered a synthetic protein to improve catalyst utilization in electrochemical H_2 production, and demonstrated proof-of-concept that designed proteins can impact electrode structure using traditional catalyst inks. The engineered protein has multifunctional properties that arrange platinum catalyst and ionomer into a highly organized network. The resulting phase separation in Nafion® ionomer was investigated using atomic force microscopy in an ink containing engineered protein and compared to an ink without proteins. An improvement in phase separation in ionomer yields a promising model for advanced materials development for H_2 production.

538775 Dual Set-Point Cascade Control for Water Management in Methanol Fuel Cells

Abstract:

Dual Set-Point Cascade Control for Water Management in Methanol Fuel Cells

Zuhair A. Alyousef, Shyam P. Mudiraj, and Oscar D. Crisalle

Department of Chemical Engineering, University of Florida

Gainesville, FL32611, United States

Extended abstract submitted for presentation in the 2018 AIChE Annual Meeting

Direct methanol fuel cells (DMFCs) are highly promising candidates for powering portable devices because they have high energy density, low operational temperatures, and can be realized in a compact design [4]. Figure1 shows a schematic diagram of the DMFC system considered in this work. The device includes a mixing tank containing a diluted methanol solution, and an associated cascade feedback loop shown with standard instrumentation indicated by the symbols LT and LC to denote level transmitter and level controller, and a fan that operates as the final control element. The area between the anode flow channel (AFC) and the cathode flow channel (CFC) represents a stack of fuel cells, including the polymer exchange membrane (PEM).

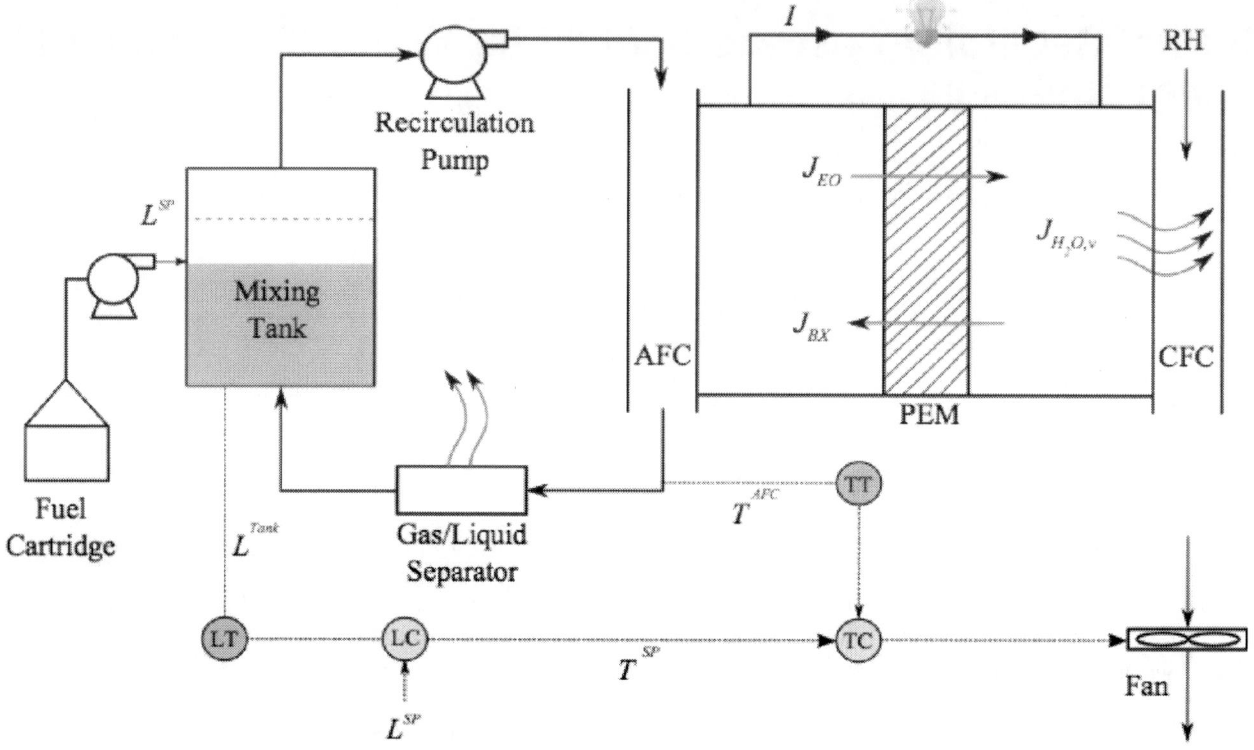

Figure 1: DMFC system.

An electrochemical reaction takes place at the anode region inside the fuel-cell stack, where the methanol and water are consumed via the electro-oxidation reaction

$$CH_3OH + H_2O \longrightarrow CO_2 + 6H^+ + 6e^- \quad (1)$$

while on the cathode side oxygen is consumed by the electro-reduction reaction

$$6H^+ + 6e^- + \frac{3}{2}O_2 \longrightarrow 3H_2O \quad (2)$$

The protons produced in (1) migrate from the anode to the cathode, passing through the PEM. In addition, some methanol transports through the membrane from the anode to the cathode, where it undergoes the undesirable crossover reaction

$$CH_3OH + \frac{3}{2}O_2 \longrightarrow 2H_2O + CO_2 \quad (3)$$

The water balance in the membrane is affected by liquid water that moves through membrane pores from the anode to the cathode via an electro-osmotic drag flux J_{EO}, while at the same time some of the liquid water produced at the cathode returns to the anode in the form of a hydraulic-pressure back-convection flux J_{BX} through small laser-drilled holes across the membrane. Water vapor losses are incurred through venting into the CFC (J) and at the gas/liquid separator.

H_2O,v

The control instrumentation shown in Figure 1 implements a water management strategy involving two proportional-integral (PI) controllers configured in a cascade mode. The master controller is denoted as LC and the slave as TC. An original control goal was to deploy the cascade scheme for the purpose of maintaining the liquid-level in the mixing tank at a fixed set point L^{SP} . The control objective of the original design is to regulate the liquid-level in the tank by adjusting the fuel cell stack temperature using the fan [2]. This scheme succeeds in maintaining the liquid level at the set point, though this is accomplished by continuously manipulating the temperature of the stack.

The master control law is described by

$$T^{SP} = \bar{T} + K_P e(t) + K_I \int_0^t e(\tau) d\tau \quad (4)$$

where T is the bias temperature (the desired temperature of operation), $e(t) = L^{SP}(t) - L(t)$ is the error of the liquid-level L(t), while K_P and K_I are the proportional and integral gains, respectively. This original control scheme is in principle able to keep the mixing tank level at set point. However, this is done at the cost of constantly adjusting the set point T^{SP} for the fuel cell stack. Hence, the operating temperature is a function of the liquid level, a scenario that is not conducive to optimal power generation.

In this work, we introduce a new cascade control policy, with the goal of decoupling, at least temporarily, the liquid-level control from the control of the temperature of the stack. This allows the introduction of periods of operation where the stack temperature can be maintained at an optimal constant value. The novelty introduced is the modification of the master-controller equation (4) in the form

$$T^{SP} = \bar{T} + f(t) \left[K_p e(t) + K_I \int_0^t e(\tau) d\tau \right] \quad (5)$$

where the switching function

$$f(t) = \begin{cases} 0 & \text{PIOFF} \\ 1 & \text{PI ON} \end{cases} \quad (6)$$

is a binary signal. The state $f(t) = 0$ implements a regime where the master PI controller is on the off mode, and produces the constant output $T^{SP} = T$. On the other hand, the state $f(t) = 1$ returns the control law (5) to its previous state (4).

Figure 2 illustrates the algorithm for switching f(t) between its two states. We refer to this technique as a dual-threshold scheme. The figure shows an upper threshold r, a lower threshold r, and a set point for liquid-level $r(t) = L^{SP}(t)$. Suppose that, as indicated in the figure, during an initial period between t_0 and t_2, the liquid-level is in a rising mode. In such case $f(t) = 0$ and the master-controller PI of equation (5) is in the off (manual) mode, as indicated by the label "OFF" in the figure. During the period of time up to the instant t_2 the output of (5) is the constant temperature target $T^{SP} = T$, which allows the operator to specify a desirable stack operating temperature T that remains constant up to instant t_2. When the liquid level reaches the upper threshold, at instant t_2, the switching function changes its state to $f(t) = 1$, and placing the controller into the automatic (on) mode. The label “ON” at instant t_3 illustrates the sate of the controller. The controller remains in the on mode until it brings the liquid level into a dead-band zone labeled 2b around the liquid level set point r. The dead-band is intended to make the algorithm less sensitive to measurement noise, and the parameter b is adjusted based on the standard-deviation of the noise. Once inside the dead-band, the switching function changes to $f(t) = 0$, placing the controller back into manual mode. This is illustrated in the figure by the label “OFF” shown at instant t_4. Hence, the dual-threshold algorithm allows the fuel cell stack to operate at desirable temperature, from the beginning to the instant t_2, and then again for another period that begins at instant t_4. An analogous logic applies to the case where the liquid level reaches the lower threshold r.

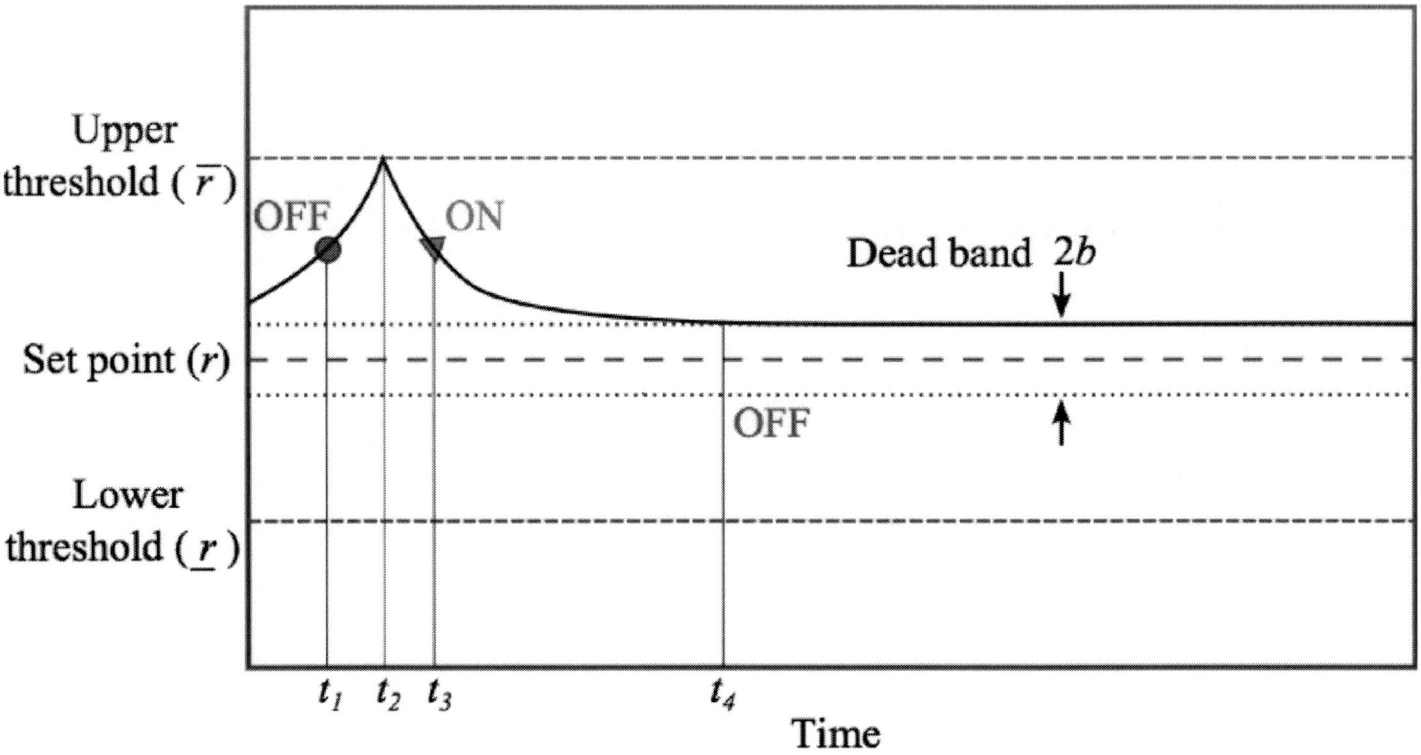

Figure 2: Schematic diagram of hypothetical example explaining the decoupled cascade control algorithm.

In this study we compare and contrast the performance of the original cascade scheme involving Equation (4) with the switching dual-threshold scheme described by Equation (5). The analysis is done in a simulation environment using the MATLAB and Simulink software packages. The dynamics of DMFC model used in the simulation has been validated experimentally by our research group [1, 3]. The simulation involves introducing a step disturbance in the relative humidity RH, and recording the closed loop responses of the two controllers. The desired temperature of operation is set at T = 60°C.

Simulation results for the original cascade control scheme show that the liquid-level of the tank can be regulated successfully to the set point, but during the entire period of operation the target fuel cell temperature of 60°C is never realized since it fluctuates between 51 and 57°C. In contrast, the dual-threshold cascade control results show that the desired constant temperature of operation is realized for an extended period, essentially for most simulation duration.

Figure 3 shows the results of electrical generation under both control schemes. The original cascade control algorithm (denoted by SC in the graph) generates electrical current and power outputs with fluctuating behavior. In contrast, the dual-threshold controller values are constant for long periods of time. The improved performance of the dual-threshold algorithm is a consequence of the fact that the current and power outputs of the DMFC are strong functions of temperature. During the simulation length the controller succeeds in keeping the fuel cell stack at the optimal target temperature of 60°C. Furthermore, the total energy produced under dual-threshold control is higher than the original controller by approximately 11%. Another advantage of the dual-threshold control scheme is that by operating at the optimal temperature for long (though intermittent) periods of time, the DMFCs capability of avoiding flooding and dry-out conditions in the membrane improves. The liquid-level in the tank is adjusted satisfactorily by the dual-threshold controller during the relatively short period that it in the "ON" state. The original cascade scheme is effective in keeping the mixing tank liquid level near the set point, a result that is obtained at the expense of reduced electrical performance.

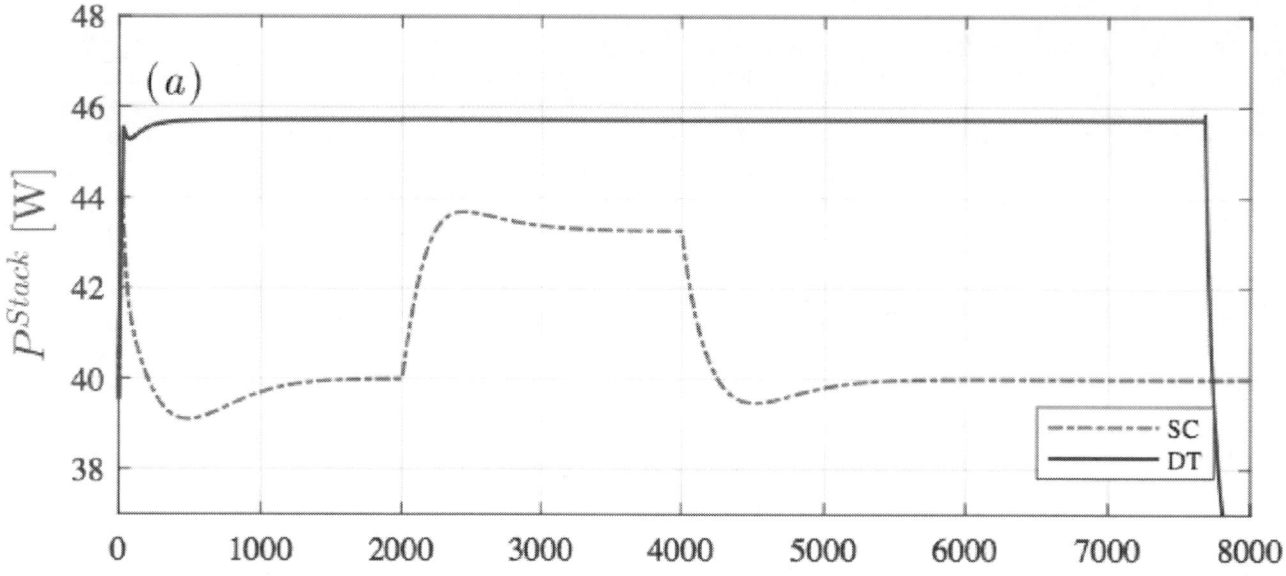

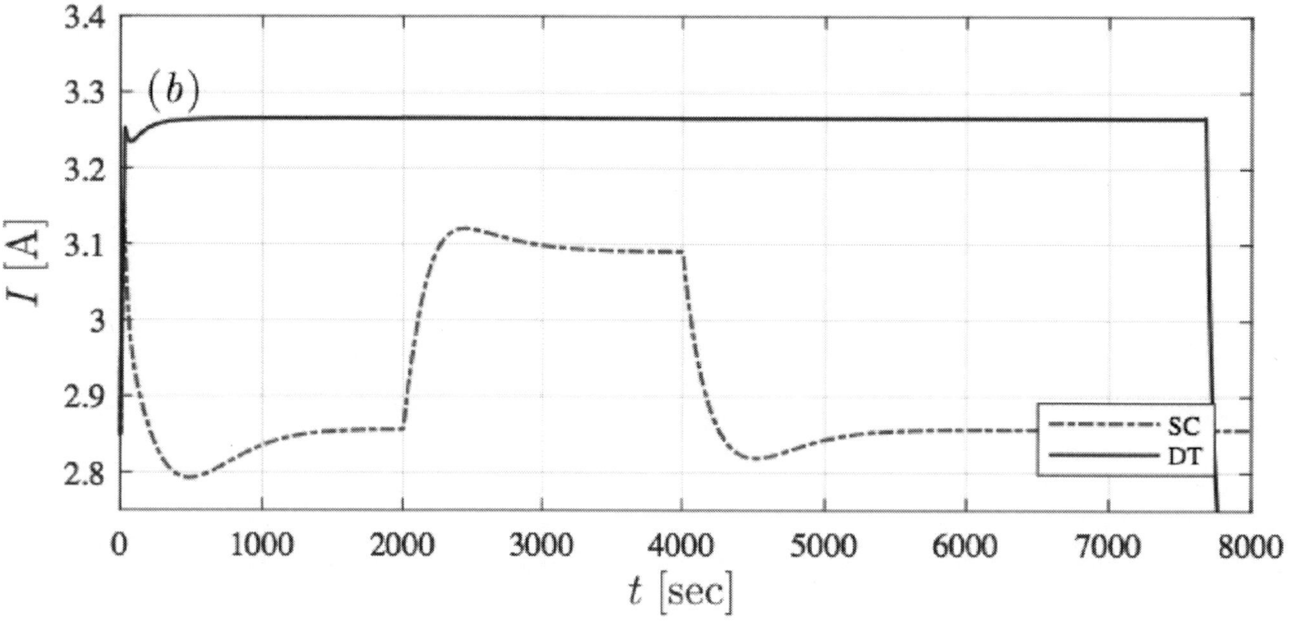

Figure 3: Simulation results showing the electrical current (I) and electrical power $\left(P^{Stack}\right)$ as a function of time for the original master controller (SC) and the dual-threshold controller (DT).

References

[1] M.A.R. Biswas, S.P. Mudiraj, W.E. Lear, and O.D. Crisalle. Systematic approach for modeling methanol mass transport on the anode side of direct methanol fuel cells. International Journal of Hydrogen Energy, 2014.

[2] Shyam P. Mudiraj. Advanced Model-based Control Design for Direct Methanol Fuel Cell Systems. phdthesis, University of Florida, April 2015.

[3] S.P. Mudiraj, M.A.R. Biswas, W.E. Lear, and O.D. Crisalle. Comprehensive mass transport modeling technique for the cathode side of an open-cathode direct methanol fuel cell. International Journal of Hydrogen Energy, 40:8137–8159, 2015.

[4] F. Zenith and U. Krewer. Modelling, dynamics and control of a portable dmfc system. Journal of Process Control, 20:630–642, 2010.

527451 A Computational Study on H_2 Absorption in a Porous Framework Structure or: The Curse of the Exponential Function

Abstract:

In the context of hydrogen storage, we have proposed a covalent organic framework based on adamantane units that can be decorated with Sc atoms, which in return act as anchor sites for adsorbing H_2 molecules.[1] We will first briefly describe the idea behind the design of the framework. Afterwards, we will illustrate our simple thermodynamics model used to determine macroscopic observables like the H_2 coverage rate from adsorption energies computed at DFT-level. The model approximates all nuclear degrees of freedom of the framework and the adsorbed H_2 as harmonic vibrations. We will demonstrate the importance of the thermodynamic model for the actual evaluation of the performance of a H_2 storage material, as the amount of adsorbed hydrogen is highly dependent on the operating conditions.

In a second part, we will discuss the error of the calculated results. We will show that a small deviation in the Gibbs free energy of H_2 adsorption can lead to a large error in the hydrogen adsorption rate. The reason for this is the exponential dependence of the adsorption constant on the adsorption energy. In addition, we will also discuss potential error sources of the Gibbs free energy of adsorption, among others the treatment of (hindered) rotation or translation modes as harmonic vibrations.

References

1. A. Deshmukh, T.N.M. Le, C.-c. Chiu, J.-L. Kuo,. *submitted.*

529190 Towards an Ultrasensitive Method for the Quantification of Metal Ions Using Surface-Enhanced Infrared Absorption Spectroscopy

Abstract:

We have developed an ultrasensitive method for the quantification of metal ions using attenuated total reflectance surface enhanced infrared absorption spectroscopy (ATR-SEIRAS). The detection limit of metal ions is as low as 10 ppq, which is two orders of magnitude lower than that of currently used ICP-MS/electrochemical methods and thus provides a more sensitive approach to understanding of the structure-stability relationships of catalysts at atomic level.

527041 Two Stage Stochastic Programming Modeling Approach for Energy Storage System Operation Under Uncertainty

Abstract:

Employing renewable energy sources has become an irreversible trend in the current energy systems of various sectors including process industries. Their decision-makers should cope with often occurring mismatches between varying energy supply and demand over multiple time periods. Considering the inherent variability of renewable output, it would be often witnessed that surplus energies from renewable sources is simply thrown away without being used. It is thus understandable to see the introduction of energy storage system (ESS) for the sake of the economic benefit of using the surplus again later. We can increase the penetration of renewable energies by installing the ESS. On the other hand, it is also right to point out that there is much to be explored in maximizing the potential of ESS with other entities in the energy system. In order to transform the expectation into reality, the corresponding decision-making framework should be rigorously designed. As an alternative, this paper investigates the impact of variation on the design and operation of ESS. Specifically, the design of ESS under uncertainty is transformed into a two-stage stochastic programming model. The stochastic model was formulated in terms of a mixed integer linear programming (MILP) problem. As illustrated in numerical examples, the proposed modeling approach allows us to gain the insight on renewable energy systems. More similar works are expected to establish the robust energy system in process industries.

527226 Theoretical and Experimental Investigation of the Microbial Degradation of Solitary Oil Microdroplets

Abstract:

Natural seeps and accidental releases of crude oil in the sea result in clouds of droplets that are carried away by underwater sea currents. The droplets may be created either at the sea surface during the breakup of an oil slick by sea waves, or at the seafloor during the atomization of live crude oil extruding at sufficiently high speed from a natural crack or a broken wellhead. Drifting droplet clouds disturb the established ecosystem dynamics and pose a high risk of toxic effects to many marine species. It is therefore of crucial importance to understand and quantify the factors that determine the retention of the dispersed droplets within the seawater column. Dissolution, biodegradation and aggregation with marine snow are the main physical processes that rule the fate of dispersed oil droplets in marine waters.

With regard to the biodegradation process, microbes have developed three fundamental strategies for accessing and assimilating oily substrates. Depending on their affinity for the oily phase and ability to proliferate in multicellular structures, microbes might either attach to the oil surface and directly uptake compounds from the oily phase, or grow suspended in the aqueous phase consuming solubilized oil, or form three-dimensional biofilms over the oil-water interface (Kapellos, 2017).

We have recently developed a compound particle model for the biodegradation of solitary oil microdroplets moving through a water column (Kapellos et al., 2018). The compound particle is of the core-shell type and consists of an oily core that is successively surrounded by a bioreactive skin of negligible thickness and another bioreactive shell of finite thickness. The bioreactive skin represents a thin layer of microbes that uptake oil directly from the oily phase, whereas the bioreactive shell represents a distinct biofilm phase. The new model accounts for all three modes of biodegradation: direct interfacial uptake, bioreaction in the bulk aqueous phase, and bioreaction in a biofilm formed around the droplet. A closed-form correlation has been established for the overall dissolution rate for a given particle configuration and two coupled ordinary differential equations have been derived for the evolution of the dimensions of the compound particle. Numerical analysis is currently used to extend the domain of validity of the model by relaxing key hypotheses. In this presentation, major findings of the theoretical analysis will be discussed in detail and in conjunction with preliminary results from the biodegradation of hexadecane droplets by Marinobacter sp. in a microfluidic setting.

Acknowledgements: This work has received funding from the European Unionâ€™s Horizon 2020 research and innovation programme under the Marie Sklodowska-Curie grant agreement No 741799 (project "OILY MICROCOSM").

References:

Kapellos, G.E. "Microbial strategies for oil biodegradation", In *Modeling of Microscale Transport in Biological Processes*; Becker, S.M., Ed.; Academic Press, pp. 19-39 (2017).

Kapellos, G.E., Paraskeva, C.A., Kalogerakis, N. and Doyle, P.S. "Theoretical insight into the biodegradation of solitary oil microdroplets moving through a water column", *Bioengineering*, 5:15, (2018).

530591 Modeling Faradaic Capacitive Deionization with Redox Active Porous Electrodes

Abstract:

Capacitive deionization (CDI) is an attractive technology for brackish water treatment, due to the low energy consumption at low source water salinity level compared to conventional water treatment technologies such as reverse osmosis or distillation.

It has been shown that the surface charge on the porous carbon electrodes has significant impact on CDI performance. By introducing immobilized chemical groups, the point of zero charge of carbon materials can be tuned, thus enhanced or inversed CDI performance could be obtained. Redox active electrodes have been investigated in CDI applications recently. The Faradaically induced chemical charges add to electronic charges within the electric double layers formed at the electrode surfaces to provide a much higher adsorption capacity than attained in purely capacitive processes, which is denoted as Faradaic capacitive deionization (FaCDI) here. The redox active ligand can be considered as variable chemical charges that depend on the cell voltage, as opposed to the fixed charges introduced by chemical modification.

Here, we developed a model which incorporates the effect of variable chemical charge due to general faradaic reactions of redox species immobilized on the electrode surfaces. The theory describes the variable chemical charge by the generalized Frumkin-Butler-Volmer equation across the stern layer as an extension of current electrosorption theory. As a first approximation, the model assumes the spacer channel is well-mixed with uniform electrosorption in each electrode. The model is able to predict the experimentally observed enhanced and inverted performance of CDI cells. Furthermore, the deionization performance of FaCDI cells is shown to be superior to that of CDI and enhanced-CDI systems with equilibrium adsorption capacities 50-100% higher than attained with CDI systems. By properly choosing the equilibrium potentials of the redox active material, FaCDI is able to achieve the same ion adsorption capacity at smaller voltages compared to conventional CDI, which indicates the promise of obtaining higher salt adsorption capacity with lower energy input through the use of FaCDI materials.

The model enables us to optimize the operation parameters of the process, such as flow rate, applied voltage as well as the design of CDI cells, such as the surface to volume ratio of the device, in order to obtain the optimal deionization performance.

532901 Finding the Ideal Feedstock for Biodiesel Production Via Generalized Kinetic Model for Transesterification and Saponification

Abstract:

The hunt for an assortment of energy sources is just about as old as humanity's need for energy. This search has been escalated by an expanding industrialization since the 19th century. The internal combustion engine is a prime example. Even though fuels obtained from fossil sources have been a prime choice since the onset of these engines, biofuels especially biodiesel have also found some interest. This interest has particularly gained momentum in the past few decades due to rapidly increasing energy demand and depletion in fossil fuel resources.

Biodiesel is a non-toxic, biodegradable, and environmentally benign fuel that can be used in diesel engines. The most commonly used method for the production of biodiesel is transesterification. It involves the reaction of oil with an alcohol in the presence of a catalyst to form fatty acid alkyl ester (biodiesel) and glycerol. Although, biodiesel is a promising alternative fuel, its production poses significant challenges due to its higher costs compared to that of petroleum-based diesel. Typically one of the major contributors to this issue is the cost of the feedstock, which accounts for approximately 60 to 80% of the total production cost.

Even though the literature is replete with the studies about modelling and optimization of transesterification [1, 2, 3, 4], very few researchers have focused their attention on the selection of feedstock for biodiesel production. Although, a few comparative studies have been reported in the literature [2, 5], all of them focus on static optimization, thereby neglecting the kinetics of biodiesel production. Therefore, in order to address these issues, we aim to achieve the following objectives:

1. Find optimal operating conditions of the reactor for biodiesel production using different vegetable oils as feedstock.
2. Develop meta-models depicting the effect of the different vegetable oils/their blends on the properties of the biodiesel produced using these feedstocks.
3. Use the developed meta-models to find the best possible feedstock for a particular country.

The purpose of this study is to find optimal feedstock (vegetable oil blend) for a particular country based on the standard property specifications and the availability of oils in that country. To this end, we use the generalized kinetic model, developed in our previous work [6] to find optimal conditions with 4,500 different blends of six major oils *viz.*, Jatropha, Linseed, Olive, Palm, Rapeseed, and Sunflower oil as feedstock. We employ a property based objective function (considering density, viscosity, cold point, pour point, cetane number, higher heating value, lower calorific value, and cold filter plugging point of biodiesel) to solve these optimization problems. We then employ correlations from the literature [3] to calculate properties of biodiesel formed using a particular feedstock and at corresponding optimal conditions. This is followed by developing meta-models with the computed properties of biodiesel as output variables and different characteristics of the feedstocks as the input variables. We use five different approaches to characterize the feedstock *viz.* a) average chain length and the unsaturation degree of the fatty acids that constitute the oil (adopted from [7], number of dimensions (N) = 2); b) average chain length, weighted unsaturation degree calculated based on the triglyceride composition of oil (proposed in this work, N = 2); c) triglyceride composition (proposed in this work, N = 34); d) weights on the considered oils for the different blends (proposed in this work, N = 6); and e) properties of oils, *i.e.* viscosity, density, heat capacity, enthalpy of vaporization, and vapour pressure calculated based on the triglyceride composition of oil (proposed in this work, N = 5). Finally, we use the best performing meta-models to obtain optimal feedstock for production of biodiesel in a particular country.

The predictions made using the developed meta-models are in accord with those made using the original models. We quantify this agreement, by computing the Percentage Average Absolute Error (PAAE) as follows,

PAAE = (Average Absolute Error/Average Response)x100

The error values are below 1% for most of the cases with the maximum value at 2.55%, which shows excellent prediction performance. Moreover, we use a normalized error metric to compare the performance of the meta-models developed using different data sets in this study. We observed that the meta-models developed using triglyceride composition and weight distribution of oils in the blend as input variables performed superior to the other meta-models. However, owing to the large number of dimensions in the former case, it is apt to use the latter. Finally, we used the meta-models to obtain optimal feedstock blend for a particular country. We analysed different scenarios based on the standard specification imposed on the biodiesel quality, availability and type of feedstock (*i.e.* edible or non-edible). We observed that the quality of biodiesel produced can change significantly based on the considered scenarios, thereby highlighting the importance of such a study. Besides, we also observed that it is possible to produce premium quality biodiesel even with just non-edible oils. Such a result can definitely bolster the efforts for commercialization of biodiesel production as the use of non-edible oils can significantly reduce the costs associated with the biodiesel production process.

REFERENCES

[1] V. Singh, K. Solanki, M. N. Gupta, Process optimization for biodiesel production, Recent patents on biotechnology 2 (2) (2008) 130{143.

[2] J.Yang, T. Astatkie, Q. S. He, A comparative study on the effect of unsaturation degree of camelina and canola oils on the optimization of bio-diesel production, Energy Reports 2 (2016) 211-217.

[3] S. Pinzi, D. Leiva, G. Arzamendi, L. Gandia, M. Dorado, Multiple response optimization of vegetable oils fatty acid composition to improve biodiesel physical properties, Bioresource technology 102 (15) (2011) 7280-7288.

[4] R. S. Kumar, K. Sureshkumar, R. Velraj, Optimization of biodiesel production

from manilkara zapota (l.) seed oil using taguchi method, Fuel 140 (2015) 90-96.

[5] T. Issariyakul, A. K. Dalai, Comparative kinetics of transestercation for biodiesel production from palm oil and mustard oil, The Canadian Journal of Chemical Engineering 90 (2) (2012) 342-350.

[6] P. Chhabra, S. Mosbach, M. Kraft, I. A. Karimi, A generalized kinetic model for transestercation and saponification (2017).

URL https://aiche.confex.com/aiche/2017/meetingapp.cgi/Paper/500597

[7] E. G. Giakoumis, A statistical investigation of biodiesel physical and chemical properties, and their correlation with the degree of unsaturation, Renewable Energy 50 (2013) 858-878.

530255 Accelerating the Generation of Coal Power Plant Property Models

Abstract:

Power plants are complex systems that are represented by first principles or empirical models. As power plants are developed with new technologies outside of well-studied thermodynamic regions, data-driven approaches have been used to model these complex systems [5]. For example, thermophysical properties or kinetic reaction models can be explained to an arbitrary level of complexity with an empirical model. The challenge is that these process models are not well understood, and it is not always intuitive what functional forms can be utilized to represent these systems. When optimizing proposed model systems, the efficiency of power plants relies on having accurate models. To maximize efficiency as well as to maintain safe operating conditions, it is essential that complex kinetic and thermophysical relationships are represented accurately, even though determining such a representation may be computationally expensive.

In this work we address the problem of speeding up the model generation process. In particular, we look at accelerating best subset selection algorithms with the use of GPU parallel computing. Best subset selection is a regression problem that seeks to determine a small set of features that best relates a set of inputs to an output [7]. There are many exact and heuristic approaches that can be used to solve the best subset selection problem, such as branch-and-bound [4], forward and backward selection [3], exhaustive search, the lasso [7], and mixed-integer optimization [6, 2, 1]. The proposed approach speeds up the model building process, allowing us to consider a larger feature space and generate more realistic models. The proposed methodology is compared against best subset selection techniques as well as the Automated Learning of Algebraic MOdels (ALAMO) methodology [2].

References

[1] D. Bertsimas, A. King, and R. Mazumder. Best subset selection via a modern optimization lens. The Annals of Statistics, 44:813–852, 2016.

[2] A. Cozad, N. V. Sahinidis, and D. C. Miller. Automatic learning of algebraic models for optimization. AIChE Journal, 60:2211–2227, 2014.

[3] M.A. Efroymson. Multiple regression analysis. Mathematical methods for digital computers, pages 191–203, 1960.

[4] C. Gatu and EJ Kontoghiorghes. Branch-and-bound algorithms for computing the best-subset regression models. Computational and Graphical Statistics, 15:139–156, 2006.

[5] S Lu and BW Hogg. Dynamic nonlinear modelling of power plant by physical principles and neural networks: International Journal of Electrical Power & Energy Systems, 22.

[6] A. Miller. Subset selection in regression. CRC Press, 2002.

[7] R. Tibshirani. Regression shrinkage and selection via the lasso. Royal Statistical Society, Methodological:267–288, 1996.

524382 Dynamic Optimisation of Water-Injection Wells Operation for Enhanced Oil Production from a Mature Oil and Gas Field

Abstract:

Field engineers in the oil and gas industry are constantly faced with the challenge of maintaining profitability amidst several operational constraints whose effects span a wide spectrum of timescales. Given the inevitably high environmental and financial stakes associated with the exploration and production of an oil and gas prospect, there is a strong incentive to enhance hydrocarbon recovery and production via systematic, model-based optimisation approaches [1-3]. In further response to this challenge, the application of sophisticated simulation methodologies to comprehensively ocapture the reservoir behaviour, multiphase flows (in wellbores and flowlines) [4-5] and gas-oil-water separation in the processing facilities is of exceptional industrial as well as academic interest. Liquid loading in gas wells and artificial lift design considerations, reservoir pressure maintenance via water injection, gas/water coning during production from deviated wells, pressure drop and liquid holdup of multiphase mixtures in highly deviated flowlines are some of the specific features of the problem which are frequently encountered and known to affect its complexity.

Simulation of these prevalent subsurface and surface phenomena does not always guarantee an accurate prediction of the onset of these problems, let alone the trouble-free operation of an existing petroleum field. In order to tackle the shortcomings and thus improve the fidelity of the currently employed models, it is necessary to also combine robust optimisation methods with these oil and gas CFD simulations [6-7]. This combination of simulation and optimisation algorithms increases the order of complexity due to model nonlinearity, non-convexity and the presence of discrete variables. There has been an increasing number of contributions in the application of optimisation techniques to field production, and this can be attributed to the advances in the development of specialised algorithms and accompanying computational power. To date, most of these attempts have focused on fields undergoing primary production through predominantly vertical wells [1-3].

This paper advances the state of the art in this field by formulating a production optimisation problem, via simultaneous consideration of production and injection wells with both vertical and deviated (inclined/horizontal) well geometries. A field undergoing secondary production (Enhanced Oil Recovery, EOR) is considered in this study. We use multiphase flow and reservoir simulation software in order to compute flowrates and pressure drops in wellbores and flowlines of the considered production network; extra measures are taken to ensure all flowlines are free from hydrate and wax at the prevalent temperature and pressure conditions. The problem is solved as a nonlinear program (NLP), comprising an economic objective function and several constraints to ensure operational feasibility. The adopted optimisation technique yields a reasonable increase in oil production compared to the usual and direct application of a black-box simulator. This also translates to an increased NPV with the current oil price, thus demonstrating the efficiency of the proposed method as a value addition tool.

REFERENCES

[1] Gunnerud, V., Conn, A. and Foss, B., 2013. Embedding structural information in simulation-based optimization. *Comput. Chem. Eng.* **53**: 35–43.
[2] Gerogiorgis, D.I., Georgiadis, M., Bowen, G., Pantelides, C.C. and Pistikopoulos, E.N., 2006. Dynamic oil and gas production optimization via explicit reservoir simulation. *Comput. Aided Chem. Eng.*, **21**: 179–184.
[3] Tavallali, M.S., Karimi, I.A., Teo, K.M., Baxendale, D. and Ayatollahi, S., 2013. Optimal producer well placement and production planning in an oil reservoir. *Comput. Chem. Eng.*, **55**: 109–125.
[4] Epelle, E.I. and Gerogiorgis, D.I., 2017. A multiparametric CFD analysis of multiphase annular flows for oil and gas drilling applications. *Comput. Chem. Eng.*, **106**: 645–661.
[5] Epelle, E.I. and Gerogiorgis, D.I., 2018. Transient and steady state analysis of drill cuttings transport phenomena under turbulent conditions. *Chem. Eng. Res. Des.* **131**: 520-544.
[6] Gunnerud, V. and Foss, B., 2010. Oil production optimization—A piecewise linear model, solved with two decomposition strategies. *Comput. Chem. Eng.*, **34**(11): 1803–1812.
[7] Kosmidis, V.D., Perkins, J.D. and Pistikopoulos, E.N., 2004. Optimization of well oil rate allocations in petroleum fields. *Ind. Eng. Chem. Res.* **43**(14): 3513–3527.

537394 Voluntary Oil Well Ignition As a Blowout Response: Analysis of Factors Influencing Viability

Abstract:

While most off-shore drilling use conventional steel rigs, for large reservoirs in shallow waters, artificial islands make an attractive alternative for offshore drilling because it offers more space, lower environmental impact, greater personnel safety, and lower cost. Drilling from an off-shore island may allow voluntary ignition as a blowout response if the oil contains sufficient gas. For voluntary ignition to be a viable blowout response, wellhead burning â€“ that depends on a number of factors â€“ needs to be nearly 100% efficient.

In this work, we investigate wellhead burning as a planned response to oilwell blowouts. The work presents a case study of proposed wellhead burning of 880 GOR, 27° API oil, should a blowout occur. The work takes into account some of the concerns regarding possible significant oil spill on Beaufort Sea as a result of incomplete combustion of the blowout effluent.

Burning of two-phase mixtures of oil and gas is a complex phenomenon that depends on the fraction of gas in the mixture, percentage of oil being entrained as liquid droplets in the gas stream, the size of the droplets, their residence time in the flame, among other parameters. Available models and correlations for the influence of these parameters on burning efficiency are examined in view of their relevance to wellhead burning.

This work will demonstrate that for some oil wells with high enough GOR, wellhead burning could be efficient enough for to be a reasonable response to a well blowout. In the studied case, a GOR of 880 combined with high total flowrate of the mixture, strongly implies nearly complete combustion of the well effluent during a blowout. The case study helped us formulate an algorithm to examine whether any particular well could be a possible candidate for wellhead burning as a planned response to a blowout. The approach suggests determining a wellâ€™s worst case discharge (WCD), estimating fluid velocities at well exit, estimating adiabatic flame temperature, and performing an energy balance to determine if all oil drops will evaporate.

AUTHOR INDEX

AUTHOR INDEX

AUTHOR INDEX

AIChE
120 Wall Street, FL 23
New York, NY 10005-4020

ISBN 978-1-5108-7626-2

9 781510 876262

CURRAN ASSOCIATES INC.
proceedings.com